Stoddard's Mental Exercises.

THE

AMERICAN

INTELLECTUAL ARITHMETIC

DESIGNED FOR THE USE

OF

SCHOOLS AND ACADEMIES;

CONTAINING

AN EXTENSIVE COLLECTION OF PRACTICAL QUESTIONS, WITH CONCISE AND ORIGINAL METHODS OF SOLUTION, WHICH SIMPLIFY MANY OF THE MOST IMPORTANT RULES IN WRITTEN ARITHMETIC.

By JOHN F. STODDARD, A. M.,

PRINCIPAL OF THE

LANCASTER COUNTY NORMAL SCHOOL,

AUTHOR OF

"THE JUVENILE MENTAL," "THE PRACTICAL," AND "THE PHILOSOPHICAL" ARITHMETICS, "READY RECKONER," ETC.

NEW YORK:

SHELDON, BLAKEMAN & CO.

PHILADELPHIA, J. B. LIPPINCOTT & CO. BOSTON, JOHN P. JEWETT & CO. BUFFALO, PHINNEY & CO. CLEVELAND, J. B. COBB & CO. CINCINNATI, APPLEGATE & CO. DETROIT, KERR, MORLEY & CO. CHICAGO, KEENE & LEE. NASHVILLE, TOON, NELSON & CO. ST. LOUIS, E. K. WOODWARD AND KEITH & WOODS.

PREFACE.

In presenting the following work to the public, I have been prompted by no desire of *pecuniary* gain, or wish to appear in the capacity of an *author*. But having felt the necessity of a *more extended*, and *more systematically* arranged work on the subject, for the benefit of *younger* as well as *older* and *more advanced* pupils;—and, by using it in manuscript form, in my own school, for a number of years, having observed the *superior mental training* to be derived from its study, and the ease with which it enabled pupils to comprehend the more advanced branches of mathematics, I offer it to a discriminating public.

While one object has been to produce a more profound work on this important branch of study, than any with which I am acquainted; I have at the same time exercised much thought in *arranging* and *classifying* the subjects, so that they are adapted to the comprehension of very young pupils.

The rule which I have observed in preparing this work is, *Tell but* ONE *thing at a time, and* THAT *in its proper place.*

It would be laborious to point out all the particulars in which this work differs from other similar works; I shall, therefore, mention those only which occur in giving a brief exposition of its general plan, leaving the many other *differences* (which, doubtlessly, will be considered of much importance) to be found by those who study the book.

CHAPTERS FIRST, SECOND, THIRD, and FOURTH, treat respectively of *Addition*, *Subtraction*, *Multiplication*, and *Division* of simple numbers; each of which is rendered familiar by an extensive collection of practical examples. The last *Lesson* in *Chapter Second*, consists of practical questions, which combine *Addition* and *Subtraction*. The last *Lesson* in *Chapter Third*, consists of practical questions combining *Addition*, *Subtraction* and *Multiplication*. *Thus*, an intimate connection between *Lessons* and even *Chapters* is kept up through the whole work, with the exception of *Chapter Fifth*, which contains a few of the most important tables of *Weights* and *Measures*; each of which is illustrated with appropriate examples. The table of *Sterling Money*, being foreign to our currency, has been omitted.

CHAPTER SIXTH is devoted to the subject of *Fractions*, and contains *twenty* lessons, in which many *original combinations* and *concise* solutions occur.

CHAPTER SEVENTH consists of practical and intricate questions of various kinds, which require for their solution a *thorough knowledge* of the preceding *Chapters*. This *Chapter* (the greater part of which is believed to be found in no similar work) when thoroughly understood, will be of *incalculable* benefit to those who are studying, or intend to study *Algebra*.

CHAPTER EIGHTH includes *Interest*, *Discount*,

and *per cent.* of every description, in all their various modifications. The method of treating these subjects is *original;* and renders the rules in *Written Arithmetic*, under these heads (which are usually so incomprehensible to pupils) *perfectly intelligible* by reducing the *whole* to one continued train of reasoning.

It is believed that this *Chapter*, if *thoroughly* taught, cannot fail to *quicken*, *strengthen*, and *develop* the reasoning powers; bringing into exercise, as it does, nearly every *principle* taught in the twenty lessons of *Chapter Sixth*, and also the greater part of *Chapter Seventh*, it must of necessity cause the pupil to acquire the habit of *systematically classifying* his knowledge, that he may, at any time, be able to call to his aid, such portions of it as will assist in *illustrating* or *demonstrating* the subject under consideration;—a *habit*, of infinite importance to a person in every condition of life.

The mind is composed of a variety of faculties which require for their development *appropriate* and *constant* exercise. That *Intellectual Arithmetic*, properly taught, is better calculated, than any other study, to *invigorate* and *develop* these *faculties*, to produce *accurate* and *close discrimination*, and, to enable the pupil to acquire a knowledge of the *Higher Mathematics* with greater ease, *cannot* for a moment *admit* of a doubt.

J. F. STODDARD.

LIBERTY JAN. 1849.

SUGGESTIONS TO TEACHERS

To those whose *experience* is *limited*, the *Author* would beg leave to present the following suggestions in regard to the *most approved* methods of teaching this *important branch of study.*

First.—The lessons should be assigned previously to recitation, to afford the pupil an opportunity for its examination: the use of the book, during class exercise, should be *entirely prohibited.*

Secondly.—To concentrate the attention of the whole class, questions should be assigned *promiscuously*, and not in *rotation* as is too frequently done.

Thirdly.—No question should be read more than once, if done *slowly* and *distinctly;* the student should be required to *reproduce* and *solve* it without interruption, unless it be to make a *necessary criticism* or *correction.* Care should be taken that the language of the pupil be *rigidly accurate*, as to *construction* and *articulation.*

Fourthly.—It is respectfully suggested that, the particular forms given for the solution of questions be carefully adhered to, unless, *better* ones should be devised by the teacher.

J. F. S.

ARITHMETIC.

CHAPTER I.

LESSON I.

1. 2 and 1 are how many?
2. 2 and 2 are how many?
3. 2 and 3 are how many?
4. 2 and 4 are how many?
5. 2 and 5 are how many?
6. 2 and 6 are how many?
7. 2 and 7 are how many?
8. 2 and 8 are how many?
9. 2 and 9 are how many?
10. 3 and 2 are how many?
11. 3 and 3 are how many?
12. 3 and 4 are how many?
13. 3 and 5 are how many?
14. 3 and 6 are how many?
15. 3 and 7 are how many?
16. 3 and 8 are how many?
17. 3 and 9 are how many?
18. 4 and 3 are how many?
19. 4 and 4 are how many?
20. 4 and 5 are how many?
21. 4 and 6 are how many?
22. 4 and 7 are how many?
23. 4 and 8 are how many?
24. 4 and 9 are how many?

25. James killed 2 birds and John 1; how many did they both kill?

26. Gave 2 cents to Henry, and 2 cents to Harvey; how many cents did they both receive?

27. Hiram had 2 cents, and his brother gave him 3 more; how many had he then?

28. George gave me 2 apples, and Mary gave me 4; how many did they both give me?

29. A man had 2 cows, and he purchased 5 more; how many cows did he then have?

30. John's father gave him 2 oranges, and his mother gave him 6; how many did he receive in all?

31. Philo bought 2 peaches, and his brother gave him 7 more; now many has he in all?

32. Philip gave me 2 plums, and Myron gave me 8; how many did they both give me?

33. A farmer had 2 horses, and bought 9 more; how many had he then?

34. William had 3 candies, and Moses gave him 2 more; how many did he then have?

35. John had 3 apples, and I gave him 3 more, how many had he then?

36. Philip gave 3 cents for some nuts, and 4 cents for some candies; what did he pay for both?

37. I paid 3 cents for some wafers, and 5 cents for a stamp; what did they both cost me?

38. A merchant bought 3 barrels of sugar, and 6 barrels of molasses; how many barrels did he then have?

39. Ralph is 3 years old, and Edward is 7; what is the sum of their ages?

40. A lemon cost 3 cents, and a pine-apple cost 8; what was the cost of both?

41. James solved 3 questions in arithmetic, and Oliver 9; how many did they both solve?

42. If it take 4 yards of cloth for a coat, and 3 for pair of pants, how many yards will it take for both?

43. Samuel had 4 marbles, and found 4 more; how many had he then?

44. Isaac bought 4 sheets of paper, and I gave him 5 more; how many had he then?

45. If a peck of apples cost 4 cents, and a peck of pears 6 cents, what is the cost of the apples and pears together?

46. If Mary has 4 books, and her father should give her 7 more, how many would she then have?

47. William has 4 marbles in his hand, and 8 in his pocket; how many has he in all?

48. Charles walked 4 miles, and rode 9; how far did he go?

49. In a certain class there are 5 boys, and 4 girls; how many scholars are there in the class?

50. Jacob bought 6 apples, and Jeremiah bought 7; how many did they both buy?

LESSON .I.

1. 5 and 4 are how many?
2. 5 and 5 are how many?
3. 5 and 6 are how many
4. 5 and 7 are how many?
5. 5 and 8 are how many?
6. 5 and 9 are how many?
7. 6 and 5 are how many?
8. 6 and 6 are how many?
9. 6 and 7 are how many?
10. 6 and 8 are how many?
11. 6 and 9 are how many?
12. 7 and 6 are how many?
13. 7 and 7 are how many?
14. 7 and 8 are how many?
15. 7 and 9 are how many?

16. 8 and 7 are how many?
17. 8 and 8 are how many?
18. 8 and 9 are how many?
19. 8 and 5 are how many?
20. 9 and 6 are how many?
21. 9 and 8 are how many?
22. 9 and 9 are how many?
23. 9 and 10 are how many?
24. 9 and 7 are how many?
25. 9 and 11 are how many?

26. Mary has answered 5 questions correctly and 4 incorrectly; how many questions was she asked?

27. A beggar met two boys; one gave him 5 cents, and the other gave him 6 cents; how many cents did they together give him?

28. A man bought a hat for 5 dollars, and a pair of boots for 6 dollars; what was the cost of both?

29. There are 9 boys on one bench, and 8 on another; how many are there on both?

30. Maria gave her teacher 5 pinks and 7 roses; how many flowers did she give him?

31. Harry caught 5 squirrels, and Henry caught 8; how many did they both catch?

32. If we learn 5 pages this week, and 9 next; how many will we learn in two weeks?

33. Frank sold a melon for 6 cents, and an orange for 5 cents; how many cents did he receive for both?

34. John had 6 whips, and Joseph gave him 6 more; how many had he then?

35. George had 6 chestnuts, and Richard gave him 7 more; how many had he then?

36. Henry bought 6 candies, and Sarah bought 8; how many were bought by both?

37. Rebecca has 6 oranges, and Catherine has 9; how many oranges have Rebecca and Catherine?

38. A boy had 7 apples, and his father gave him 6 more; how many had he then?

39. Minerva had 7 yards of ribbon, and her mother gave her 7 more; how many yards did she then have?

40. 7 boys were sitting on one bench, and 8 on another; how many were there on both?

41. 7 boys were at play, and 9 other boys joined hem; how many were there in all?

42. If I have 8 cents in one hand, and 7 cents in the other, how many have I in both?

43. If Mary has 8 peaches, and Margaret has 9, how many have they both?

44. Sally gave 9 cents for some thread, and 7 cents for some needles; how much did the needles and thread cost her?

45. Charles has 9 marbles, and Albert has 5; how many marbles have Charles and Albert?

46. 9 birds were in a tree, and 6 were on the ground; how many were there in all?

47. Sarah gave 9 cents for some cinnamon, and 7 cents for some raisins; how many cents did they cost her?

48. George shot 9 pigeons, and James shot 8; how many did they both shoot?

49. Russel caught 7 fish, and Robert caught 5; how many did they both catch?

50. In one field there are 8 horses, and in another there are 9; how many are there in both?

LESSON III.

1. How many are 10 and 2? 10 and 3? 10 and 4? 10 and 5? 10 and 6? 10 and 7? 10 and 9? 10 and 8? 10 and 10?

2. How many are 2 and 2? 2 and 12? 2 and 22? 2 and 32? 2 and 42? 2 and 52? 2 and 62? 2 and 72? 2 and 82? 2 and 92?

3. How many are 3 and 3? 3 and 13? 3 and 23? 3 and 33? 3 and 43? 3 and 53? 3 and 63? 3 and 73? 3 and 83? 3 and 93? 96 and 4?

4. How many are 4 and 4? 4 and 14? 4 and 24? 4 and 34? 4 and 44? 4 and 54? 4 and 64? 4 and 74? 4 and 84? 4 and 94? 98 and 2?

5. How many are 5 and 5? 5 and 15? 5 and 25? 5 and 35? 5 and 45? 5 and 55? 5 and 65? 5 and 75? 5 and 85? 5 and 95?

6. How many are 6 and 6? 6 and 16? 6 and 26? 6 and 36? 6 and 46? 6 and 56? 6 and 66? 6 and 76? 6 and 86? 6 and 96?

7. How many are 7 and 7? 7 and 17? 7 and 27? 7 and 37? 7 and 47? 7 and 57? 7 and 67? 7 and 77? 7 and 87? 7 and 97?

8. How many are 8 and 8? 8 and 18? 8 and 28? 8 and 38? 8 and 48? 8 and 58? 8 and 68? 8 and 78? 8 and 88? 8 and 98?

9. How many are 9 and 9? 9 and 19? 9 and 29? 9 and 39? 9 and 49? 9 and 59? 9 and 69? 9 and 79? 9 and 89? 9 and 99?

10. How many are 10 and 11? 10 and 21? 10 and 31? 10 and 41? 10 and 51? 10 and 61? 10 and 71? 10 and 81? 10 and 91?

11. How many are 10 and 12? 10 and 22? 10 and 32? 10 and 42? 10 and 52? 10 and 62? 10 and 72? 10 and 82? 10 and 92?

12. How many are 10 and 4? 10 and 14? 10 and 24? 10 and 34? 10 and 44? 10 and 54? 10 and 64? 10 and 74? 10 and 84? 10 and 94?

13. How many are 11 and 3? 11 and 13? 11 and 23? 11 and 33? 11 and 43? 11 and 53? 11 and 63? 11 and 73? 11 and 83? 11 and 93?

14. How many are 11 and 4? 11 and 14? 11 and 24? 11 and 34? 11 and 44? 11 and 54? 11 and 64? 11 and 74? 11 and 84? 11 and 94?

15. How many are 10 and 5? 10 and 15? 10 and

35? 10 and 45? 10 and 55? 10 and 65? 10 and 75? 10 and 85? 10 and 95? 10 and 25?

16. How many are 11 and 5? 11 and 15? 11 and 25? 11 and 35? 11 and 45? 11 and 55? 11 and 65? 11 and 75? 11 and 85? 11 and 95?

17. How many are 3 and 8? 3 and 18? 3 and 28? 3 and 38? 3 and 48? 3 and 58? 3 and 68? 3 and 78? 3 and 88? 3 and 98?

18. How many are 8 and 4? 8 and 14? 8 and 24? 8 and 34? 8 and 44? 8 and 54? 8 and 64? 8 and 74? 8 and 84? 8 and 94?

19. How many are 7 and 5? 7 and 17? 7 and 27? 7 and 37? 7 and 47? 7 and 57? 7 and.67? 7 and 77? 7 and 87? 7 and 97?

20. How many are 8 and 6? 8 and 16? 8 and 26? 8 and 36? 8 and 46? 8 and 56? 8 and 66? 8 and 76? 8 and 86? 8 and 96?

LESSON IV.

1. 8 and 9 are how many?
2. 11 and 7 are how many?
3. 10 and 9 are how many?
4. 7 and 14 are how many?
5. 6 and 12 are how many?
6. 9 and 15 are how many?
7. 11 and 18 are how many?
8. 15 and 12 are how many?
9. 14 and 13 are how many?
10. 16 and 14 are how many?
11. 21 and 12 are how many?
12. 24 and 13 are how many?
13. 25 and 15 are how many?
14. 27 and 13 are how many?
15. 23 and 17 are how many?

16. 29 and 11 are how many?
17. 30 and 20 are how many?
18. 34 and 15 are how many?
19. 32 and 18 are how many?

REMARK.—The symbol +, is called *plus;* and denotes that the quantities between which it is placed, are to be added Thus 4 + 2 shows that and 2 are to be added;—and is read, *four plus two.*

20. 33 + 44 are how many?
21. 35 + 15 are how many?
22. 36 + 12 are how many?
23. 40 + 36 are how many?
24. 40 + 29 are how many?
25. 44 + 20 are how many?
26. 48 + 32 are how many?
27. 45 + 35 are how many?
28. 4 + 8 + 6 are how many?
29. 8 + 2 + 7 are how many?
30. 10 + 7 + 3 are how many?
31. 12 + 10 + 9 are how many?
32. 15 + 12 + 6 are how many?
33. 18 + 4 + 10 are how many?
34. 24 + 16 + 12 are how many?
35. 22 + 33 + 11 are how many?
36. 15 + 16 + 2 are how many?
37. 28 + 12 + 15 are how many?
38. 46 + 24 + 19 are how many?
39. 12 + 8 + 6 + 4 are how many?
40. 24 + 10 + 6 + 12 are how many?

LESSON V.

1. Three boys, James, Joseph and Jacob, gave some money to a beggar; James gave him 6, Joseph 8, and Jacob 10 cents; how many cents did they all give him?

2. Gave 8 cents to John, 4 cents to Morgan and 2 cents to Samuel; how many cents did they all receive?

3. Henry has 3 marbles, Harvey has 10, and Harry has 7; how many marbles have they all?

4. Gave 7 nuts to one boy, 6 to another, and 7 to another; how many nuts did the three boys receive?

5. Bought a basket of strawberries for 7 cents, a basket of cherries for 4 cents, and a basket of plums for 8 cents; what did they all cost?

6. Lydia has 9 pinks, Mary 10, and Ann 7; how many pinks have they all?

7. Bought a knife for 14 cents, and a ball for 12 cents; what did they both cost?

8. Gave 18 cents for an arithmetic, 2 for a pencil, and 10 for a slate; what did they all cost?

9. James had 12 cents, and his mother gave him 13 more; how many had he then?

10. Robert shot 9 birds, Richard shot 11, and James shot 12; how many did they all shoot?

11. A boy bought a pound of butter for 14 cents, a pound of meat for 8 cents, and a bunch of lettuce for 7 cents; what was the whole cost?

12. Bought a pound of raisins for 10 cents, a pound of candies for 12 cents, and a pound of cinnamon for 15 cents; what was the whole cost?

13. John had 20 marbles, Matthew 9, and Morgan 12; how many did they all have?

14. James bought a pigeon for 9 cents, a robin for 10 cents, and a squirrel for 12 cents; what did they all cost him?

15. A lady bought some pins for 15 cents, some thread for 10 cents, and some lace for 18 cents; what did all these articles cost her?

16. A gentleman bought a hat for 6 dollars, a vest for 5 dollars, and a coat for 20 dollars; what did he pay for them all?

17. A man bought a watch for 40 dollars, a gold chain for 15 dollars, and a gold pen for 5 dollars; what did he pay for these three articles?

18. Jackson gave 25 cents to his sister, and 23 cents to his mother; how many cents did he give away?

19. Bought a barrel of flour for 7 dollars, a barrel of pork for 12 dollars, and a barrel of fish for 11 dollars; what was the whole cost?

20. Bought a horse for 60 dollars, a cow for 20 dollars, and a colt for 25 dollars; what did they all cost?

21. If your father should give you 12 cents, your mother 14 cents, and your sister 4 cents; how many cents would you then have?

22. A boy spent 11 cents for confectionary, 9 cents for a ball, and 5 cents for a top; what did they all cost him?

23. A man bought some butter for 57 dollars, and some molasses for 23 dollars; what was the cost of both?

24. A boy travelled 17 miles one day, and 23 the next; how far did he travel in the two days?

25. A lady bought a hat for 7 dollars, a dress for 9 dollars, and a gold watch for 60 dollars; what was the whole cost?

26. A mechanic sold a wagon for 57 dollars, and a sleigh for 43 dollars; what did he receive for both?

27. A boy saw 24 pigeons on one tree, and 36 on another; how many did he see in all?

28. In a certain recitation 21 questions were answered correctly, and 9 incorrectly; how many questions were asked during the recitation?

29. Gave 87 dollars for a chaise, 2 dollars for a whip, and 11 dollars for a Buffalo-robe; what was the whole cost?

30. If a horse is worth 60 dollars, and a sleigh 75 dollars, what is the value of both?

31. Matthew is 15 years old, Morgan is 7, and Martin is 10; what is the sum of their ages?

32. A man bought a load of hay for 7 dollars, a load of rye for 36 dollars, and a load of wheat for 57 dollars; what was the whole cost?

33. A man is 48 years old, and his wife is 32 years old; what is the sum of their ages?

34. A farmer bought a horse for 60 dollars, and a yoke of oxen for 75 dollars; what did the horse and oxen cost him?

35. John gave 11 candies to his brother, 9 to his sister, and kept 12 himself; how many had he at first?

36. Simeon hoed 12 rows of corn, Simon 15, James 13, and John 11; how many rows did they together hoe?

37. A merchant sold 30 barrels of flour one week, 37 the next week, and 33 the following week; how many barrels did he sell during the three weeks?

38. A merchant sold a barrel of sugar for 5 dollars, a barrel of rum for 15 dollars, and a hogshead of molasses for 23 dollars; what did he receive for all these articles?

39. A man bought a firkin of butter for 9 dollars, a keg of molasses for 7 dollars, a box of cheese for 4 dollars, and a box of raisins for 5 dollars; what was the whole cost?

40. A lady bought a silk dress for 18 dollars, a muff for 12 dollars, a shawl for 17 dollars, and a pair of gloves for 1 dollar: the whole cost is required.

CHAPTER II.

LESSON I.

1. If I have 3 apples, and give 1 of them to Richard, how many shall I have left?

SOLUTION.—If I have 3 apples and give 1 of them away, there will be remaining the difference between 3 and 1 which is 2. Therefore I have 2 left.

2. William had 4 chestnuts, and gave 1 to his brother; how many had he left?

3. Martha had 5 books, and on her way to school she lost 1 of them; how many had she left?

4. Cornelia had 6 apples and gave 1 to her brother; how many had she left?

5. Rachel had 10 pins and lost 1 of them; how many had she left?

6. Martha had 12 pears and gave 2 to Elizabeth; how many had she left?

7. If you had 5 candies and should give 2 of them away, how many would you have left?

8. James had 6 apples and gave 2 away; how many had he remaining?

9. Cornelia says she has learned 12 letters to-day, but has forgotten 3 of them; how many does she remember?

10. James had 10 marbles and lost 3; how many had he left?

11. Mary found 9 roses on her bush, and picked off 4 of them; how many remained on the bush?

12. 4 less 2 are how many?

13. 5 less 3 are how many?

14. 7 less 3 are how many?

15. 9 less 4 are how many?

16. 9 less 3 are how many?

17. 9 less 7 are how many?
18. 9 less 5 are how many?
19. 11 less 4 are how many?
20. 10 less 4 are how many?
21. 11 less 5 are how many?
22. 14 less 4 are how many?
23. 8 less 5 are how many?
24. 13 less 3 are how many?
25. 14 less 5 are how many?
26. 17 less 7 are how many?
27. 15 less 5 are how many?
28. 13 less 10 are how many?
29. 23 less 3 are how many?
30. 27 less 7 are how many?

REMARK.—The symbol —, is called *minus;* and denotes that the quantity on the right of it is to be taken from the quantity on the left. Thus, 6 — 4 shows that 4 is to be subtracted from 6;—and is read, *six minus four.*

The symbol =, means *equal to,* and when placed between two quantities it denotes that they are equal to each other. Thus, 4 + 2 = 6, is read 4 plus 2, or 4 added to 2 is equal to 6.

31. 8 — 5 are how many?
32. 9 — 7 are how many?
33. 10 — 8 are how many?
34. 11 — 8 are how many?
35. 12 — 6 are how many?
36. 13 — 8 are how many?
37. 14 — 8 are how many?
38. 18 — 8 are how many?
39. 22 — 12 are how many?
40. 24 — 14 are how many?
41. 12 — 4 are how many?
42. 28 — 8 are how many?
43. 20 — 5 are how many?
44. 20 — 8 are how many?
45. 20 — 9 are how many?
46. 20 — 7 are how many?
47. 20 — 10 are how many?
48. 20 — 15 are how many?
49. 24 — 10 are how many?

50. 25 — 10 are how many?
51. 26 — 10 are how many?
52. 28 — 10 are how many?
53. 27 — 10 are how many?
54. 29 — 10 are how many?
55. 32 — 10 are how many?
56. 34 — 10 are how many?
57. 36 — 10 are how many?
58. 35 — 10 are how many?
59. 37 — 10 are how many?
60. 38 — 10 are how many?
61. 39 — 10 are how many?
62. 47 — 10 are how many?
63. 40 — 12 are how many?
64. 42 — 20 are how many?
65. 45 — 20 are how many?
66. 46 — 20 are how many?
67. 47 — 20 are how many?
68. 47 — 37 are how many?
69. 49 — 19 are how many?
70. 52 — 22 are how many?
71. 54 — 34 are how many?
72. 56 — 46 are how many?
73. 57 — 27 are how many?
74. 58 — 48 are how many?
75. 62 — 30 are how many?
76. 65 — 40 are how many?
77. 68 — 48 are how many?
78. 74 — 34 are how many?

79. Gave 7 cents for a spool of thread, and 4 cents for a lemon; how much more did the thread cost than the lemon?

80. Paid 18 cents for a pound of butter, and 8 cents for a pound of meat; how much more was paid for the butter than for the meat?

81. James bought 18 candies, and gave John 7 of them; how many had he left?

82. Sold a quantity of wool for 27 dollars, and received in payment a barrel of flour worth 5 dollars; how many dollars remain due?

83. James has 27 marbles, and John has 17; how many more has James than John?

84. Harry is 15 years old, and Henry is 9 years old; how many years older is Harry than Henry?

85. A teacher being asked how many scholars he had, answered that he usually had 37, but at present he had only 27; how many were absent?

86. A man purchased a watch for 37 dollars, but found he had only 24 dollars with him; how much must he borrow in order to pay the balance?

87. Mr. A has 94 sheep, and B has 44; how many more sheep has A than B?

88. Morgan gave 23 cents for some cake, and 14 cents for some cinnamon; how much more did the cake cost than the cinnamon?

89. Michael had 29 cents and lost 14; how many had he left?

90. In a certain recitation 47 questions were asked, and 9 of them were answered incorrectly; how many were correctly answered?

91. A man sold 23 sheep from a flock consisting of 93; how many sheep remained?

92. Mr. B bought a horse for 35 dollars, and sold it for 46 dollars; how much did he gain?

93. A cow was bought for 25 dollars and sold for 19 dollars; how much was the loss?

94. A merchant bought a quantity of goods for 95 dollars, but being damaged was obliged to sell them for 80 dollars; how much did he lose?

95. From a vessel containing 57 gallons, 27 gallons leaked out; how much remained in?

96. A merchant bought a quantity of silk for 47 dollars, and sold it for 67 dollars; how much did he gain by the bargain?

97. A butcher has 57 sheep and 44 lambs; how many more sheep has he than lambs?

98. Paid 97 dollars for a quantity of sugar, and 43 dollars for some molasses; how much more did the ugar cost than the molasses?

LESSON II.

CHAPTER FIRST COMBINED WITH THE PRECEDING LESSON.

1. A boy has 7 chestnuts in one hand, and 4 in the other; how many more has he in one hand than in the other, and how many in both?

2. Bought a barrel of fish for 8 dollars, and some quinces for 3 dollars; how much more did the fish cost than the quinces? What was the cost of both?

3. Gave 15 dollars for a cow and 6 dollars for a sheep; how much more was given for the cow than for the sheep? What was given for both?

4. Phineas gave 50 cents for a grammar, and 25 cents for an arithmetic; what was the cost of both? How much did one cost more than the other?

5. Paid 15 dollars for a barrel of rum, and 6 dollars for a barrel of flour; what was the cost of both; and how much more did the rum cost than the flour?

6. Sold a firkin of butter for 10 dollars, a keg of cheese for 5 dollars, and received in payment a barrel of flour worth 6 dollars; how much remains due?

7. James gave 12 cents for oranges, 15 cents for cake, and had 13 cents remaining; how much had he at first?

8. Mary bought a comb for 10 cents, a spool ot thread for 12 cents, and a paper of needles for 8 cents;—she handed the clerk 37 cents; how much change ought she to receive?

9. A man sold a cow for 20 dollars, a calf for 4 dollars, and a sheep for 3 dollars; and in payment received a wagon worth 17 dollars;—how much remains due?

10. A lady bought a ribbon for 24 cents, some tape for 8 cents, and some thread for 12 cents,—she had only 60 cents; how much remained after she paid for these articles?

11. Stephen, at a game of marbles, won 4 and lost 6, and then had only 8 remaining; how many had he at first?

12. Sampson having 9 apples, gave 4 to his mother and 3 to his sister;—for his generosity his father gave him 13 more; how many had he then?

13. A man bought some cloth for 12 dollars, and sold it for 18 dollars; what was his gain?

14. A farmer bought a horse for 63 dollars and exchanged it for a yoke of oxen,—these he sold for 8[illegible] dollars; what did he gain by the operation?

15. A man bought a yoke of oxen for 97 dollars, their services amounted to 40 dollars, and their keeping to 13 dollars,—he then sold them for 80 dollars; did he gain or lose, and how much?

16. A box of raisins was bought for 3 dollars, a firkin of butter for 15 dollars, and were both sold for 20 dollars; how much was gained?

17. A farmer sold a cow for 29 dollars, which was 5 dollars more than she cost; what did she cost?

18. A drover bought some sheep for 40 dollars, some cattle for 130 dollars, and sold them all for 200 dollars; what was his gain?

19. A jeweler bought a watch for 20 dollars, a chain for 10 dollars, a key for 2 dollars, and sold them all for 42 dollars; how much did he gain by the bargain?

20. 24 + 12 + 9 are how many?

21. 10 + 30 + 15 are how many?

22. $14 + 16 + 11$ are how many?
23. $36 + 9 - 12$ are how many?
24. $38 + 22 - 15$ are how many?
25. $43 + 37 - 20$ are how many?
26. $13 + 26 - 25$ are how many?
27. $44 - 22 + 10 - 12$ are how many?
28. $27 + 23 - 20 + 2$ are how many?
29. $15 + 25 - 30 + 15$ are how many?
30. $20 + 40 - 30 + 10$ are how many?

31. A boy bought a ball for 6 cents; for what must he sell it to gain 4 cents?

32. A merchant bought a hogshead of molasses for 47 dollars, and paid 3 dollars for cartage; for what must he sell it to gain 12 dollars?

33. A grocer bought a hogshead of sugar for 30 dollars; for what must he sell it to gain 18 dollars by the bargain?

34. A drover bought sheep as follows: of one man he bought 24, of another 8, and of another 22,—he then sold 20 of them; how many remained unsold?

35. A watch cost 40 dollars; how must it be sold to gain 13 dollars?

36. Four boys bought a melon; one gave 3 cents, another 4, another 8, and the other 6; what did the melon cost?

37. Mary bought 16 candies at one shop, and 13 at another,—on her way home she ate 11 of them; how many had she left?

38. Matthew had 9 nuts, Mary gave him 10 more, and John gave him enough to make his number 39; how many did John give him?

39. A farmer had 25 sheep in one field and 15 in another,—he then bought enough to make his number 56; how many did he buy?

40. John has 34 marbles, and Albert 25; how many have they both; and how many more has the one than the other?

CHAPTER III.

LESSON I.

1. Two times 1 are how many?
2. Two times 2 are how many?
3. Two times 3 are how many?
4. Two times 4 are how many?
5. Two times 5 are how many?
6. Two times 6 are how many?
7. Two times 7 are how many?
8. Two times 8 are how many?
9. Two times 9 are how many?
10. Two times 10 are how many?
11. Two times 11 are how many?
12. Two times 12 are how many?
13. What cost 2 oranges, at 3 cents a piece?

SOLUTION.—If 1 orange cost 3 cents, 2 oranges will cost two times 3 cents; which are 6 cents.

14. What cost 2 peaches, at 2 cents a piece?
15. What cost 2 apples, at 3 cents a piece?
16. What cost 2 pine-apples, at 8 cents a piece?
17. What cost 2 pounds of meat, at 5 cents a pound?
18. What cost 2 pounds of cinnamon, at 11 cents a pound?
19. What cost 2 pounds of raisins, at 12 cents a pound?
20. What cost 2 citrons, at 10 cents a piece?
21. What cost 2 quarts of cherries, at 9 cents a quart?
22. What cost 2 lemons, at 4 cents a piece?
23. Three times 2 are how many?
24. Three times 3 are how many?
25. Three times 4 are how many?

26. Three times 5 are how many?
27. Three times 6 are how many?
28. Three times 7 are how many?
29. Three times 8 are how many?
30. Three times 9 are how many?
31. Three times 10 are how many?
32. Three times 11 are how many?
33. Three times 12 are how many?
34. Four times 3 are how many?
35. Four times 4 are how many?
36. Four times 5 are how many?
37. Four times 6 are how many?
38. Four times 7 are how many?
39. Four times 8 are how many?
40. Four times 9 are how many?
41. Four times 10 are how many?
42. Four times 11 are how many?
43. Four times 12 are how many?
44. What cost 3 quarts of cherries, at 6 cents a quart?
45. What cost 3 lead pencils, at 5 cents a piece?
46. What cost 3 quarts of milk, at 4 cents a quart?
47. What cost 3 yards of ribbon, at 7 cents a yard?
48. What cost 4 quarts of chestnuts, at 6 cents a quart?
49. What cost 4 yards of edging, at 5 cents a yard?
50. What cost 3 ounces of snuff, at 8 cents an unce?
51. What cost 4 ounces of cinnamon, at 7 cents an ounce?
52. What cost 3 pounds of cheese, at 10 cents a pound?
53. What cost 4 sheets of wadding, at 8 cents a sheet?
54. What cost 3 yards of calico, at 11 cents a yard?

55. What cost 4 skeins of silk, at 9 cents a skein?
56. What cost 3 yards of ribbon, at 12 cents a yard?
57. What cost 4 pounds of starch, at 12 cents a pound?
58. What cost 4 candlesticks, at 11 cents a piece
59. What cost 4 tops, at 10 cents a piece?
60. What cost 5 apples, at 4 cents a piece?
61. Five times 6 are how many?
62. Five times 7 are how many?
63. Five times 9 are how many?
64. Five times 8 are how many?
65. Five times 5 are how many?
66. Five times 10 are how many!
67. Five times 12 are how many?
68. Five times 11 are how many?
69. Six times 6 are how many?
70. Six times 8 are how many?
71. Six times 7 are how many?
72. Six times 10 are how many?
73. Six times 9 are how many?
74. Six times 12 are how many?
75. Six times 11 are how many?
76. Seven times 6 are how many?
77. Seven times 8 are how many?
78. Seven times 7 are how many?
79. Seven times 10 are how many?
80. Seven times 9 are how many?
81 Seven times 12 are how many?
82 Seven times 11 are how many?
83. What cost 5 barrels of flour, at 6 dollars barrel?
84. What cost 5 bushels of potatoes, at 5 dimes a bushel?
85. What cost 6 primers, at 6 cents a piece?
86. What cost 5 barrels of fish, at 7 dollars a barrel?

87. What cost 6 pounds of mutton, at 7 cents a pound?

88. What cost 5 barrels of sugar, at 12 dollars a barrel?

89. What cost 6 pounds of sturgeon, at 10 cents a pound?

90. What cost 6 pounds of almonds, at 12 cents a pound?

91. What cost 5 barrels of pork, at 10 dollars a barrel?

92. What cost 6 pounds of candles, at 9 cents a pound?

93. What cost 5 coats, at 9 dollars a piece?

94. What cost 6 handkerchiefs, at 11 cents a piece?

95. What cost 6 inkstands, at 8 cents a piece?

96. What cost 7 lamps, at 9 dimes a piece?

97. What cost 7 plows, at 8 dollars a piece?

98. What cost 7 boxes of caps, at 10 cents a box?

99. What cost 7 quires of paper, at 12 cents a quire?

100. What cost 7 letter-folders, at 11 cents a piece?

101. Eight times 8 are how many?

102. Eight times 10 are how many?

103. Nine times 8 are how many?

104. Eight times 7 are how many?

105. Nine times 9 are how many?

106. Eight times 9 are how many?

107. Nine times 11 are how many?

108. Eight times 12 are how many?

109. Nine times 10 are how many?

110. Eight times 11 are how many?

111. Nine times 12 are how many?

112. What cost 9 bunches of roses at 9 cents a bunch?

113. What cost 8 pen-knives, at 12 cents a piece?

114. What cost 9 bunches of grapes, at 12 cents a bunch?

115. What cost 11 yards of calico, at 11 cents a yard?

116. What cost 10 balls of cotton, at 12 cents a ball?

117. What cost 11 pounds of ginger, at 12 cents a pound?

118. What cost 10 blocks of tape, at 8 cents block?

119. What cost 12 yards of cloth, at 12 dimes a yard?

120. What cost 13 pair of boots, at 4 dollars a pair?

LESSON II.

CHAPTERS FIRST AND SECOND COMBINED WITH THE PRECEDING LESSON.

1. At 7 cents a piece, what will 9 pine-apples cost?

2. If the postage on 1 letter is 5 cents, what will be the postage on 8 letters?

3. If it require 8 yards of calico to make 1 dress, how many yards will it require to make 7 dresses?

4. If John obtain 2 credit-marks in 1 day, how many will he have in 15 days?

5. A man hired a horse to ride, and was to give 5 cents for every mile he rode;—he rode 12 miles; how much must he pay?

6. Margaret's cloak contains 7 yards of merino, worth 9 dimes a yard; what was the value of her cloak?

7. If a stage-coach go nine miles in an hour, how far will it go in 7 hours?

8. At 2 dollars a week, what will 20 weeks board amount to?

9. The fare by railroad from Albany to Boston is 5 dollars for 1 person; what will it be for a family of 9 persons?

10. Helen had 8 rose bushes, and there were 7 roses on each; how many roses had she in all?

11. At 3 dimes a gallon, what will 15 gallons of molasses cost?

12. There are 10 rows of trees in an orchard, and 12 trees in each row; how many trees are there in the orchard?

13. A traveller meeting 13 beggars, gave to each of them 3 dimes; how many dimes did he give to them all?

14. A woman bought 11 yards of cloth and paid for it with butter,—giving 9 pounds, for a yard; how many pounds of butter did it take to pay for the cloth? How much did the cloth cost providing butter was worth 10 cents a pound?

15. In a certain corn field there are 24 rows, and 30 hills in each row; how many hills in the field?

16. What will 40 steel pens cost, at 2 cents a piece?

17. What will 8 pair of snuffers cost, at 3 dimes a pair?

18. When 2 dimes are paid for 1 duck, what will be the cost of 8 ducks? of 10 ducks? of 12 ducks?

19. When hay is worth 8 dollars a ton, what is the value of 2 tons? 4 tons? 3 tons? 7 tons? 5 tons? 10 tons? 12 tons?

20. At 2 dimes a piece, what will 4 books cost? 6 books? 8 books? 10 books? 12 books? 11 books? 7 books?

21. If 5 cents will buy 1 primer, what will be the cost of 4 primers? of 6? of 9? of 8? of 10?

22. 6 plates at 5 dimes a piece, will cost how much?

23. At 5 dimes a piece, what will 4 handkerchiefs cost? 6? 8? 10?

24. At 6 dimes a piece, what will 2 geese cost? 4? 5? 8? 10? 12? 9? 7?

25. At 12 cents a piece, what will 3 candlesticks cost? What will 6? 5? 8? 9? 10? 7?

26. If I pay 5 cents for riding 1 mile, how much must I pay for riding 7 miles? 8 miles? 9? 6? 10? 12?

27. At 7 cents a yard, what will 5 yards of ribbon cost? 6 yards? 8 yards? 9 yards? 10 yards? 12 yards?

28. If a tooth-brush cost 18 cents, what will 4 cost?

29. 9 turkeys will cost how much, at 8 dimes a piece?

30. At 14 cents a quire, what will 2 quires of paper cost? 3 quires? 4 quires? 5 quires?

31. What will 7 pictures cost, at 5 cents a piece? at 6 cents a piece? at 8 cents a piece? at 10 cents a piece?

32. What cost 8 bosoms, at 6 dimes a piece? at 10 dimes a piece?

33. At 10 dimes a piece, what will 4 caps cost? 5? 6? 8? 9?

34. At 40 cents a day, what will 2 day's work amount to? 5 days work?

35. If 1 paper of tobacco cost 6 cents, what will 3 papers cost? 5 papers? 8 papers? 12 papers?

36. At 7 dollars a hundred, what will 4 hundred feet of cedar boards cost? 9 hundred feet? 10 hundred feet?

37. If 1 bushel of wheat cost 60 cents, what will bushels cost? 4 bushels? 5 bushels?

38. What cost 8 muffs at 5 dollars each?

39. What cost 10 lead pencils, at 5 cents each?

40. What cost 11 boxes of cheese, at 4 dollars a box? at 5 dollars a box? at 8 dollars a box?

41. What cost 12 barrels of pork, at 5 dollars a barrel? at 8 dollars? at 9 dollars? at 10 dollars?

42. What cost 9 tons of hay, at 13 dollars a ton?

43. James is 9 years old, and his father is 4 times as old as he; how old is his father?

44. Jane's frock contains 7 yards of silk, worth 8 dimes a yard; what was the value of the silk? Providing the making cost 2 dollars; what was the cost of her dress?

45. If a barrel of flour will serve 12 men 8 days, how long will it serve 1 man?

46. If I spend 8 dollars in a month, and earn 12 dollars, how much shall I have at the end of 12 months?

47. If I earn 12 dollars a month, and pay 25 cents a week for washing, and 2 dollars a week for board, how much will I have at the end of 40 weeks (10 months)?

48, If I buy 9 tons of hay, at 12 dollars a ton, and sell 6 of them at 15 dollars a ton, and the 3 remaining tons at 10 dollars a ton, how much shall I gain by the operation?

49. Bought 11 yards of broad cloth at 4 dollars a yard, but being damaged I was obliged to lose 18 dollars by the sale of it; what did I receive for it?

50. If I buy 12 barrels of pork at 8 dollars a barrel, and sell it all for 108 dollars, how much shall I gain by so doing?

51. A man bought a horse for 80 dollars, paid 2 dollars a week for his keeping, and received 4 dollars a week for his work;—at the expiration of 10 weeks he sold him for 70 dollars; how much did he gain by the operation?

52. For what must I sell 4 barrels of wheat which cost me 8 dollars a barrel, to gain 8 dollars?

53. What is the cost of 9 cows, at 25 dollars each?

54. Providing a hunter should kill 5 pigeons and wound 4 every shot; how many would he kill and wound by shooting 8 times?

55. If a man travel 29 miles in a day, how many will he travel in 6 days?

56. 8 months' wages will come to how much, at 18 dollars a month?

57. If 10 men eat 18 pounds of butter in 1 week, how long would it last 1 man?

58. If 80 dollars will pay for 4 dinners for 20 men, how many dinners would it buy for 1 man?

59. Bought 3 yards of cloth for a coat, at 7 dollars a yard, the buttons and cord cost 2 dollars, buckram and wadding 1 dollar,—paid for making 6 dollars; for what must it be sold to gain 5 dollars?

60. If 17 men can do a piece of work in 9 days, how many days would it take 1 man to perform the same work?

61. Two men start from the same place, and travel in opposite directions;—one at the rate of 7 miles an hour, the other 9 miles an hour; how far apart will they be in 2 hours?

62. Two men start from the same place, and travel the same way;—one at the rate of 3 miles an hour, the other 8 miles an hour; how far apart will they be at the end of 8 hours?

63. Two men are 50 miles apart, and approach each other;—one at the rate of 2 miles an hour, the other 3 miles an hour; how far apart will they be at the end of 5 hours?

64. If 1 orange is worth 4 apples, how many apples must be given for 13 oranges?

65. A man earned 80 cents a day, and paid 50 cents a day for his board and washing; how much had he left at the expiration of 6 days?

66. Jane bought 4 yards of silk at 2 dollars a yard, 3 shawls at 10 dollars each, and some delain for 10 dollars;—she paid 5 ten dollar bills; how much ought she to receive back?

67. Mary bought 5 yards of silk at 8 dimes a yard,

8 yards of linen at 9 dimes a yard; how many yards did she buy, and what did they cost?

68. In a certain school there are 12 girls, and 3 times as many boys, lacking 8; how many boys in the school, and how many boys and girls together?

69. John has 7 books, and Mary has 4 times as many, lacking 18; how many has Mary, and how many have they both?

70. Albert has 9 marbles, Aaron 3 times as many, lacking 7, and Amos has twice as many as both, lacking 8; how many have they each, and how many have they together?

71. Perry worked for Elisha 4 days, at 6 dimes a day;—Elisha gave him 7 bushels of corn at 3 dimes a bushel; how much was then due Perry?

72. A merchant bought 25 pounds of sugar for 125 cents, and sold 15 pounds of it at 6 cents a pound, and the remaining 10 pounds at 4 cents a pound; how much did he gain by so doing?

73. If the interest on 1 dollar for a year is 6 cents, what would be the interest on 13 dollars for the same time?

74. What will 27 pounds of beef cost, at 4 cents a pound?

75. When beef is 5 cents a pound, and pork 9 cents; how much more will 9 pounds of pork cost than 9 pounds of beef?

76. Mary bought 35 quarts of milk, and on her way home spilled 4 times 2 quarts, lacking 3 quarts; how many quarts had she remaining?

77. Henry is 4 feet in height and John is 5; and 5 times the sum of their heights, considered as a number, is equal to their father's age + 15 years. Required, the father's age.

78. If an orange cost 2 cents, a lemon twice as much, and a melon 4 times as much as the orange and

lemon together, lacking 14 cents, how much more will 3 melons cost than 3 oranges and 3 lemons?

79. James has 9 walnuts, John twice as many lacking 8, and Joseph twice as many as both James and John + 7; how many has each, and how many have hey all?

80. If an apple cost 2 cents, an orange three times as much lacking 4 cents, and a pine-apple three times as much as the apple and orange + 5 cents, what will be the cost of all three?

CHAPTER IV.

LESSON I.

1. 4 are how many times 2?

SOLUTION.—4 are as many times 2 as 2 is contained in 4, which are two times.

2. 6 are how many times 2?
3. 8 are how many times 2?
4. 10 are how many times 2?
5. 12 are how many times 2?
6. 14 are how many times 2?
7. 6 are how many times 3?
8. 9 are how many times 3?
9. 12 are how many times 3?
10. 15 are how many times 3?
11. 18 are how many times 3?
12. 21 are how many times 3?
13. 24 are how many times 3?
14. 16 are how many times 2?
15. 18 are how many times 2?
16. 20 are how many times 2?
17. 22 are how many times 2?
18. 24 are how many times 2?
19. 26 are how many times 2?

20. 28 are how many times 4?

21. At 2 cents a piece, how many apples can you buy for 4 cents?

SOLUTION.—If for 2 cents I can buy 1 apple, for 4 cents I can buy as many apples as 2 is contained in 4, which are two times. Therefore, at 2 cents a piece, for 4 cents I can buy 2 apples.

REMARK.—The following *solution* is preferred to the above, if the pupils are acquainted with *Fractions*.

SOLUTION.—If for 2 cents I can buy 1 apple for 1 cent I can buy $\frac{1}{2}$ of an apple; and for 4 cents, 4 times $\frac{1}{2}$, which are $\frac{4}{2}$ or 2 apples.

22. At 2 cents a piece, how many oranges can I buy for 6 cents?

23. At 2 cents a piece, how many peaches can be bought for 8 cents?

24. At 3 dimes a yard, how many yards of calico can be bought for 12 dimes?

25. At 3 cents a piece, how many lemons can be bought for 9 cents?

26. At 2 cents a yard, how many yards of tape can be bought for 10 cents?

27. At 2 dimes a bushel, how many bushels of apples may be had for 12 dimes?

28. How many pounds of ginger, at 2 dimes a pound may be had for 14 dimes?

29. How many baskets of strawberries, at 3 cents a basket can be had for 15 cents?

30. For 16 dollars; how many yards of cloth can be had, at 2 dollars a yard?

31. For 18 apples; how many oranges can be bought, at the rate of 2 apples for 1 orange?

32. How many primers, at 2 cents a piece, can b bought for 24 cents?

33. How many barrels of flour, at 2 dollars a barrel, can be bought for 20 dollars?

34. For 22 dollars; how may sheep may be bought, at 2 dollars a piece?

35. How many melons may be had for 18 dimes, at 3 dimes a piece?

36. At 3 cents a piece, how many tops may be had for 6 cents?

37. If 1 peck of beans cost 3 dimes, how many pecks can be bought for 21 dimes?

38. At 3 cents a mile, how many miles can I ride for 24 cents?

39. How many bushels of rye, at 4 dimes a bushel, may be bought for 12 dimes?

40. How many books, at 4 dimes each, can be bought for 20 dimes?

41. 8 are how many times 4?

42. 12 are how many times 4?

43. 16 are how many times 4?

44. 10 are how many times 5?

45. 15 are how many times 5?

46. 20 are how many times 5?

47. 28 are how many times 7?

48. 32 are how many times 4?

49. 30 are how many times 5?

50. 35 are how many times 5?

51. 36 are how many times 4?

52. 40 are how many times 5?

53. 44 are how many times 4?

54. 30 are how many times 6?

55. 28 are how many times 7?

56. 24 are how many times 8?

57. 36 are how many times 9?

58. 24 are how many times 12?

59. 35 are how many times 7?

60. 38 are how many times 2?

61. At 5 dimes each, how many turkeys can be had for 25 dimes?

62. If the wages of 1 day is 4 dimes, what will be the wages for 9 days?

63. How many days will a man be required to work for 12 dimes, if he receive 4 dimes a day?

64. If a boy spends 5 cents a day, how many days will it take him to spend 15 cents?

65. A boy had 20 marbles, and divided them equally among his 5 brothers; how many did each receive?

66. A boy divided 28 cents equally among 4 poor women; how many cents did each receive?

67. A farmer gave 4 of his laborers 32 bushels of corn; how many bushels did each receive?

68. If 5 quarts of molasses cost 30 cents, what will 1 quart cost?

69. At 5 cents a yard, how many yards of ribbon may be had for 35 cents? how many for 50 cents?

70. How many pine-apples at 8 cents each, can be obtained for 40 cents? for 56 cents?

71. If a man travel 45 miles in 9 hours, how many miles does he travel in 1 hour?

72. If a man travel 5 miles in an hour, how many hours will it take him to travel 40 miles?

73. How many yards of cloth at 4 dollars a yard, can you buy for 32 dollars?

74. In a certain orchard there are 48 trees standing in rows, and 6 trees in each row; how many rows are there in the orchard?

75. For 56 dollars, how many barrels of pork can be bought, at 8 dollars a barrel?

76. If a man can travel 6 miles in an hour, how long will it take him to travel 42 miles?

77. How many yards of cloth at 4 dollars a yard, can you buy for 36 dollars?

78. A butcher gave 39 dollars for sheep, at the rate of 3 dollars a head; how many sheep did he buy?

79. 45 dollars were given for 9 barrels of flour; how much was it a barrel?

80. How long would it take to travel 72 miles, at the rate of 3 miles an hour?

LESSON II.

1. 20 are how many times 2? 4? 10?
2. 22 are how many times 11?
3. 24 are how many times 3? 4? 2?
4. 25 are how many times 5?
5. 28 are how many times 2? 7?
6. 30 are how many times 2? 3? 5?
7. 32 are how many times 2? 4? 16?
8. 34 are how many times 17?
9. 40 are how many times 2? 4? 5? 8?
10. 44 are how many times 2? 11?
11. 46 are how many times 23?
12. 48 are how many times 2? 3? 4? 6?
13. 50 are how many times 2? 10?
14. 56 are how many times 2? 7?
15. 57 are how many times 3?
16. 60 are how many times 2? 3? 4? 5? 6?
17. 64 are how many times 2? 4? 8?
18. 66 are how many times 2? 3? 6?
19. 68 are how many times 2? 4?
20. 70 are how many times 10? 2?
21. 72 are how many times 2? 4? 6? 8?
22. 5 are how many times 2, and how many remaining?

REMARK.—Whenever there is a remainder, it may be mentioned simply as a remainder.

23. 7 are how many times 2?
24. 17 are how many times 4? 2? 5?
25. 18 are how many times 6? 4? 2?
26. 34 are how many times 4? 6? 5? 2?
27. 25 are how many times 5? 4? 2? 3?
28. 16 are how many times 9? 4? 8? 7?
29. 32 are how many times 7? 5? 6?
30. 63 are how many times 9? 4? 5? 6?
31. 74 are how many times 2? 4? 6? 7?

32. 80 are how many times 2 ? 3 ? 4 ? 5 ? 6 ? 7 ? 8 ?
33. 84 are how many times 2 ? 3 ? 4 ? 5 ? 6 ? 7 ? 8 ?
34. 88 are how many times 2 ? 4 ? 3 ? 6 ? 8 ?
35. 37 are how many times 2 ? 6 ? 4 ?
36 27 are how many times 3 ? 4 ? 8 ?
37. 15 are how many times 4 ? 6 ? 7 ? 8 ?
38. 29 are how many times 2 ? 3 ? 4 ? 5 ? 6 ? 7 ?
39. 90 are how many times 2 ? 4 ? 6 ? 8 ? 9 ? 11 ?
40. 144 are how many times 2 ? 4 ? 6 ? 8 ? 12 ?

LESSON III.

1. At 2 cents each, how many lemons can you get for 14 cents ?

2. How many boxes of cheese, at 4 dollars a box, may be had for 12 dollars ?

3. If one hundred pounds of hay cost 3 dollars, how many hundred may be had for 15 dollars ?

4. If one barrel of flour support 20 persons one week, how many persons will it support 4 weeks ?

5. If 1 man can ride 1 mile for 4 cents, how far can 2 men ride for 80 cents ?

6. If 10 men accomplish a certain piece of work in 2 days, how long will it take 1 man to do the same ?

7. If 3 yards of cloth make one coat, how many coats will 18 yards make ?

8. If I receive 12 dollars interest in one year, in how many years will I receive 144 dollars interest ?

9. A man traveled 7 miles in 1 hour, at the same rate, how long would it take him to travel 63 miles ?

10. If 1 cow cost 13 dollars, how many cows may be had for 65 dollars ?

11. How many pens can you buy for 27 cents, if 1 pen cost 3 cents ?

12. If 8 apples are worth 40 chestnuts, how many chestnuts is 1 apple worth?

13. How many cents is 1 lemon worth, if 12 lem ons are worth 48 cents?

14. What will 1 cord of wood cost, if 20 cords cost 40 dollars?

15. If 1 pound of beef cost 7 cents, what will 19 pounds cost?

16. For 147 cents, how many pounds of sugar can be bought, at 7 cents a pound?

17. If 9 yards of cloth cost 53 dollars, how must it be sold a yard to gain 10 dollars?

18. If 7 barrels of flour cost 38 dollars, and were sold, at 7 dollars a barrel, what was the gain?

19. How many peaches, at 4 cents each, may be bought for 96 cents?

20. How many yards of cloth, at 4 dollars a yard, can be bought for 116 dollars?

21. How many oranges, at 3 cents each, must be given for 18 lemons worth 4 cents each?

22. If 15 sheep cost 75 dollars, what will 1 sheep cost?

23. Which will cost the most, 4 barrels of wheat flour at 9 dollars a barrel, or 12 barrels of corn at 4 dollars a barrel, and how much?

24. How many barrels of beef, at 3 dollars a barrel, can be bought for 54 dollars?

25. How many pounds of fish, at 5 cents a pound, may be had for 95 cents?

26. At 7 cents a pound, how many pounds of lead may be had for 84 cents?

27. How long will it require to travel 105 miles, at the rate of 5 miles an hour?

28. A person divided 72 peaches equally among 6 boys; how many did each receive?

29. 148 marbles were divided equally among some

boys; how many boys were there, providing each boy received 2 marbles?

30. How many pounds of butter, at 14 cents a pound, can be bought for 28 apples, at 3 cents each?

31. At 7 cents a bottle, how many bottles of ink can you buy for 14 sheets of paper, at 2 cents a sheet?

32. In how many days can 1 man do as much as 7 men in 8 days?

33. In how many days can 2 men do as much work as 6 men can in 3 days?

34. In how many days can 4 men earn as much as 8 men can in 6 days?

35. In how many days can 15 men earn as much as 3 men can in 25 days?

36. In how many months will 6 horses eat as much as 18 horses will in 5 months?

37. How many men can, in 7 days, earn as much as 28 men in 4 days?

38. In 10 days 6 men will earn as much as how many men in 5 days?

39. How many yards of cloth, at 4 dollars a yard, may be had for 4 sets of chairs, at 12 dollars a set?

40. A farmer gave 13 barrels of flour, worth 4 dollars a barrel, for 26 yards of cloth; how much was the cloth a yard?

LESSON IV.

CHAPTERS FIRST, SECOND, THIRD AND FOURTH COMBINED.

1. 4 times 6 are how many times 2?

SOLUTION.—4 times 6 are 24. 24 are as many times 2 as 2 is contained in 24, which are 12 times.

2. 4 times 9 are how many times 3?

3. 4 times 8 are how many times 2?

4. 4 times 10 are how many times 5?
5. 4 times 12 are how many times 6?
6. 4 times 14 are how many times 7?
7. 5 times 9 are how many times 15?
8. 5 times 8 are how many times 4?
9. 5 times 12 are how many times 15? 6?
10. 6 times 7 are how many times 2?
11. 6 times 8 are how many times 12? 3?
12. 4 times 6 are how many times 8?
13. 7 times 15 are how many times 5?
14. 12 times 7 are how many times 21?
15. 8 times 7 are how many times 4?
16. How many times 12 are 9 times 4?
17. How many times 10 are 5 times 4?
18. How many times 9 are 3 times 21?
19. How many times 5 are 7 times 15?
20. How many times 9 are 3 times 36?
21. How many times 12 are 9 times 13?
22. How many times 21 are 9 times 14?
23. How many times 7 are 3 times 14?
24. How many times 5 are 8 times 10?
25. How many times 5 are 10 times 6?
26. How many times 5 are 6 times 15?
27. 10 times 4, + 2 are how many times 7? 2?
28. 8 times 9, — 2 are how many times 5?
29. 12 times 8, — 8 are how many times 2?
30. 26 times 11, — 6 are how many times 28?
31. 7 times 8, + 4 are how many times 12?
32. 6 times 7, + 4 are how many times 2?
33. 5 times 8, + 8 are how many times 6?
34. 4 times 9, — 4 are how many times 8?
35. 9 times 5, — 3 are how many times 7?
36. 7 times 12, — 14 are how many times 5?
37. 4 times 15, + 4 are how many times 8?
38. 6 times 7, + 3 are how many times 9?
39. 11 times 13, — 3 are how many times 7?
40. 11 times 15, + 5 are how many times 17?

LESSON V.

PRACTICAL QUESTIONS COMBINING CHAPTERS THIRD AND FOURTH.

1. If 2 apples cost 4 cents, what will 3 apples cost?

SOLUTION.—If 2 apples cost 4 cents, 1 apple will cost 1 half of 4 cents, which is 2 cents. If 1 apple cost 2 cents, 3 apples will cost 3 times 2 cents, which are 6 cents.

2. If 2 pears cost 16 cents, what will 5 pears cost?

3. If 4 quinces cost 12 cents, what will 3 quinces cost?

4. If 6 oranges cost 18 cents, what will 9 oranges cost?

5. If 7 peaches cost 21 cents, what will 9 peaches cost?

6. If 4 lemons cost 24 cents, what will 7 lemons cost?

7. If 3 yards of tape cost 18 cents, what will 6 yards cost?

8. If 7 hair-brushes cost 28 dimes, what will 6 hair-brushes cost?

9. If 9 yards of muslin cost 108 cents, what will 7 yards cost?

10. If 11 books cost 44 dimes, what will be the cost of 7 of the same books?

11. If 12 ink-stands cost 96 cents, what will 2 cost?

12. If 10 lead-pencils cost 30 cents, what will be the cost of 7? of 9? of 2?

13. What will 13 yards of silk cost, if 5 yards cost 45 dimes?

14. If a man travel 15 miles in 3 hours, how far at this rate, can he travel in 9 hours? 5 hours? 7 hours?

15. If the cartage of a load of plaster 20 miles cost 4 dollars, how far could it be carried for 12 dollars?

16. How many pair of pantaloons can be cut out of 21 yards of cloth, if 3 pair can be cut out of 9 yards of the same kind of cloth?

17. What will 30 pounds of sugar cost, if 6 pounds cost 42 cents?

18. What will 18 pounds of veal cost, if 6 pound cost 42 cents?

19. What will 75 pounds of pork cost, if 9 pound cost 72 cents?

20. What will 20 weeks' board come to, if 7 weeks board cost 14 dollars?

21. What will be the wages for 1 year, if 4 months wages amount to 48 dollars?

22. What will be the cost of 25 bushels of apples, if 13 bushels cost 260 cents?

23. What will 14 pounds of cheese cost, if 6 pounds cost 54 cents?

24. If 7 quarts of milk cost 35 cents, what will 36 quarts cost?

25. If 4 men can do a certain piece of work in 12 days, in how many days will 3 men do the same work?

26. I gave for a quantity of cotton 72 dollars, and sold it for 12 yards of cloth; what did the cloth cost me a yard?

27. Gave 15 pounds of sugar for 5 pounds of butter; how much did the butter cost a pound, providing 8 pounds of sugar were worth 56 cents?

28. If 4 chestnuts are worth 8 walnuts, how many walnuts are 27 chestnuts worth?

29. If 7 yards of broadcloth are worth 35 dollars, how many boxes of butter, at 3 dollars a box, would 9 yards of this cloth buy?

30. A man bought 4 barrels of flour for 20 dollars, and gave 3 of them for cider, at 3 dollars a barrel; how many barrels of cider did he get?

31. A man bought 14 barrels of cider for 56 dollars, and gave 5 barrels of it for cloth, at 2 dollars a yard; how many yards did he receive?

32. A merchant, having 15 yards of cloth worth 75

dollars, gave 10 of them for pork, worth 10 dollars a barrel; how many barrels did he receive?

33. When 9 bushels of rye were worth 45 dimes, 12 bushels were given for 15 yards of cloth; what did the cloth cost a yard?

34. If 35 yards of cloth cost 140 dollars, what will 95 yards of the same cloth cost?

35. Two boys are 32 rods apart, and both running in the same direction, the hindermost boy gains on the other 4 rods each minute; in how many minutes will he overtake him?

36. How many boxes will be required to contain 56 bushels, providing each box contains 8 bushels?

37. How many barrels of onions, at 3 dollars a barrel, must be given for 21 boxes of raisins, at 2 dollars a box?

38. A farmer bought 9 yards of cloth, at 4 dollars a yard, and paid for it with cider, at 3 dollars a barrel; how many barrels did it take?

39. How long would it take a man to lay up 24 dollars, if he save 2 dollars a week?

40. A farmer hired a laborer and agreed to give him 6 dollars for every 3 days' work; how much did he receive a week, allowing 6 working days in a week? how much a month, allowing 4 weeks to the month?

41. If 4 oranges are worth 12 cents, how many oranges must be given for 6 pine-apples, worth 12 cents each?

42. How many yards of cloth, at 2 dollars a yard, can be bought for 4 reams of paper, at 5 dollars a ream?

43. 6 men bought a horse for 80 dollars, and paid 2 dollars a week for keeping him;—at the end of 10 weeks they sold him for 82 dollars; what did each man lose?

44. If 2 apples are worth 1 orange, and 2 oranges

are worth 1 lemon; how many lemons can be bought for 48 apples?

45. If 5 oranges are worth 1 pine-apple, and 2 pine-apples are worth 1 melon; how many oranges may be bought for 4 melons?

46. A fox is 80 rods before a hound, and the hound gains 5 rods on the fox every 10 minutes; in how many minutes will the fox be caught?

47. If 7 men can do a certain job of work in 12 days, in how many days could 21 men do the same work?

48. In how many days can 9 men do as much work as 7 men can in 18 days?

49. How many men can in 10 days perform the same amount of work, that 8 men can in 5 days?

50. Bought 5 firkins of butter for 35 dollars; how must it be sold to gain 10 dollars? what is the gain on each firkin?

CHAPTER V.

LESSON I.

TABLE OF UNITED STATES CURRENCY.

10 Mills	make	1 Cent,	marked	*c.*
10 Cents	"	1 Dime,	"	*d.*
10 Dimes	"	1 Dollar,	"	$.
10 Dollars	"	1 Eagle,	"	*E.*

1. How many mills in 1 cent? 3 cents? 4 cents? 5 cents?

2. How many cents in 1 dime? 2 dimes? 4 dimes? 5 dimes? 6 dimes? 9 dimes? 10 dimes?

3. How many dimes in 1 dollar? 2 dollars? 4 dollars? 3 dollars? 5 dollars? 9 dollars? 10 dollars?

4. How many eagles in 400 dimes?

5. How many dimes in 1 eagle and $4?

6. How many dollars in 40 dimes? 90 dimes? 80 dimes? 70 dimes? 68 dimes? 75 dimes?

7. How many cents in 40 mills? 60 mills? 50 mills? 80 mills? 48 mills? 55 mills? 47 mills?

8. If 3 yards of muslin cost 6 dimes, how many yards can be bought for 1 dollar?

9. How many pounds of pepper can you buy for 1 eagle, if 12 pounds cost 6 dimes?

10. How many pounds of candies can you buy for 14 eagles, if 10 pounds cost 140 cents?

LESSON II.

Table of Troy Weight.

24 Grains	make 1 Pennyweight,	marked	*pwt.*
20 Pennweights	" 1 Ounce,	"	*oz.*
12 Ounces	" 1 Pound,	"	*lb.*

1. How many pennyweights in 240 grains?

2. How many pennyweights in 4 ounces? 5 ounces? 6 ounces?

3. How many ounces in 1 pound? 3 pounds? 5 pounds? 8 pounds?

4. In 24 ounces how many pounds? in 48 ounces? in 36 ounces? in 60 ounces? in 84 ounces?

5. How many ounces in 20 pennyweights? in 40? in 60? in 70?

6. If 7 grains of gold cost 168 dimes, how much will 10 pennyweights cost?

7. What will 1 pennyweight of gold cost, if 2 grains cost 18 dimes?

8. How many pennyweights in 4 oz. and 6 pennyweights?

9. How many pounds in 48 pwt.?

10. How many grains in 2 oz., 2 pwt. and 2 gr.?

LESSON III.

TABLE OF AVOIRDUPOIS WEIGHT.

16 Drams	make	1 Ounce,	marked	*oz.*
16 Ounces	"	1 Pound,	"	*lb.*
25 Pounds	"	1 Quarter,	"	*qr.*
4 Quarters	"	1 Hundred-weight,	"	*cwt.*
20 Hundred-weight	"	1 Ton,	"	*T.*

1. How many drams in 2 ounces? in 4? in 10?
2. How many ounces in 2 pounds? in 4? in 8?
3. How many quarters in 100 pounds? in 400 pounds?
4. How many pounds in 3 quarters? in 2? in 7?
5. How many quarters in 2 hundred-weight? in 3?
6. In 8 qr. how many hundred-weight?
7. If 30 lb. of hay cost 6 dimes, what will 3 qr. cost?
8. What will 2 tons of iron cost, if 1 pound cost 1 dime?
9. What will 40 tons of hay cost, at 2 dimes a quarter?
10. What will 2 cwt. of sugar cost, at a dime a pound?

LESSON IV.

TABLE OF CLOTH MEASURE.

2¼ Inches	make	1 Nail,	marked	*na.*
4 Nails	"	1 Quarter of a yard,	"	*qr.*
4 Quarters	"	1 Yard,	"	*yd.*
3 Quarters	"	1 Ell Flemish,	"	*E. Fl.*
5 Quarters	"	1 Ell English,	"	*E. E.*
6 Quarters	"	1 Ell French,	"	*E. Fr.*

1. In 4 yds. 3 qr., how many quarters?
2. In 7 yds. 2 qr., how many Ells French?
3. In 8 yds. 3 qr., how many Ells Flemish?

4. In 8 Ells English, how many yards?

5. In 7 Ells Flemish, how many yards and quarters?

6. In 3 Ells French, how many yards and quarters?

7. In 4 Ells Fr. and 8 Ells Fl., how many yards?

8. In 1 qr., how many inches?

9. What will 6 E. E. of cloth cost, if 6 nails cost 48 cents?

10. What will 3 E. E. + 2 E. Fr. of cloth cost, if 3 nails cost 12 cents?

LESSON V.

TABLE OF WINE MEASURE.

4 Gills	make	1 Pint,	marked	*pt.*
2 Pints	"	1 Quart,	"	*qt.*
4 Quarts	"	1 Gallon,	"	*gal.*
42 Gallons	"	1 Tierce,	"	*tier.*
63 Gallons	"	1 Hogshead,	"	*hhd.*
2 Hogsheads	"	1 Pipe,	"	*pi.*
2 Pipes	"	1 Tun,	"	*tun.*

1. How many gills in 3 pints?

2. In 3 qts., how many gills?

3. In 12 gallons, how many pints?

4. What will 5 gallons of rum cost, if 3 gills cost 5 cents?

5. How many pints in 2 pipes?

6. A merchant bought a hogshead of molasses for 20 dollars, and sold it, at the rate of 15 cents for 3 pints; how much did he gain by the bargain?

7. How much will a gallon of wine cost, if 7 gills cost 21 cents?

8. In 1 tierce, how many pints?

9. In 4 quarts and 2 pints, how many gills?

10. In 1 tun, how many gills?

LESSON VI.

TABLE OF DRY MEASURE.

2 Pints	make	1 Quart,	marked	*qt.*
8 Quarts	"	1 Peck,	"	*pk.*
4 Pecks	"	1 Bushel,	"	*bu.*

1. In 1 peck, how many pints?
2. 2 pecks will fill how many pint measures?
3. In 3 pecks and 3 quarts, how many pints?
4. In 1 bushel and 3 pecks, how many quarts?
5. In 1 bushel, how many quarts? how many pints?
6. If 8 pints of nuts cost 24 cents, what will 3 pecks cost at the same rate?
7. A market-woman bought 4 quarts of strawberries for 29 cents, and sold them, at 5 cents a pint; how much did she gain?
8. A person sold 2 bushels and 1 peck of currants, at 2 cents a pint, and in payment received 1 bushel of gooseberries, at 4 cents a pint; how much remains due?
9. What will 5 quarts of wheat cost, if 1 bushel cost 128 cents?
10. A farmer sold 1 bu., 3 pk. and 1 pt. of clover-seed, at 640 cents a bushel, and in payment received 1 bu., 2 pk. and 3 quarts of grass-seed, at 320 cents a bushel; how much remains due?

LESSON VII.

TABLE OF TIME.

60 Seconds	make	1 Minute,	marked	*m.*
60 Minutes	"	1 Hour,	"	*h.*
24 Hours	"	1 Day,	"	*d.*
7 Days	"	1 Week,	"	*w.*
4 Weeks	"	1 Month,	"	*mo.*
12 Calendar months	"	1 Year,	"	*y.*
52 Weeks	"	1 Year,	"	*y.*

The following table exhibits the names of the months, and the number of days in each.

Season		Names.	Days.
Winter.	1st month	January,	31
	2d "	February,	28, in leap year 29
Spring.	3d "	March,	31
	4th "	April,	30
	5th "	May,	31
Summer.	6th "	June,	30
	7th "	July,	31
	8th "	August,	31
Autumn.	9th "	September,	30
	10th "	October,	31
	11th "	November,	30
Winter.	12th "	December,	31

☞ *In our calculations on interest we shall reckon* 30 *days to the month, and* 12 *months to the year, although not strictly accurate.*

1. In 2 hours how many seconds ?
2. In 2 weeks and 5 days, how many days ?
3. In 48 hours, how many days ?
4. 7200 seconds, how many hours ?
5. How many hours in a week ?
6. In 1 day, 12 hours and 10 minutes; how many minutes ?
7. How many hours in a month ?
8. If a boy can do a piece of work in 40 minutes, how many hours would it take him to perform 12 times as much work ?
9. If I can do a piece of work in 10 minutes, how many hours would it take to perform a piece of work 12 times as large ?
10. How many days in 3 weeks and 5 days ?

CHAPTER V

LESSON I.

1. If I cut an apple into 3 equal parts, what is 1 of these parts called? what are 2 of these parts called?

ANSWER.—1 part is called 1 third, and 2 parts are called twice 1 third, which are 2 thirds.

2. Mary had 4 dresses, and Rachel 1 half as many; how many had she?

3. James is 8 years old, and John is 1 half as old; how old is John?

4. Moses having 2 marbles, gave 1 half of them to his brother; how many had he left?

5. If you divide 6 apples equally between 2 boys, what part of them will each have?

6. What is 1 half of 6?

7. How many halves in 1?

8. If an orange cost 8 cents, and a peach 1 half as much, what is the cost of the peach?

9. James had 12 cakes, and John 1 half as many; how many had he?

10. If 3 apples cost 6 cents, what part of 6 cents will 1 apple cost?

11. What is 1 third of 6?

12. What is 1 half of 8? 10? 12? 14? 16? 18? 20?

13. If 3 quarts of strawberries cost 18 cents, what part of 18 cents will 1 quart cost? What part of 18 cents will 2 quarts cost?

14. What is 1 third of 18? 1 half of 18?

15. If 4 pounds of raisins cost 8 dimes, what part of 8 dimes will 1 pound cost? 2 pounds? 3 pounds? 4 pounds?

16. What is 1 fourth of 8?

17. What is 1 fifth of 15?

18. If 1 fifth of 15 is 3, what is 2 fifths of 15? 3 fifths? 4 fifths? 6 fifths? 8 fifths?

19. What is 1 sixth of 12?

20. If 1 sixth of 12 is 2, what is 2 sixths of 12? 3 sixths? 4 sixths? 5 sixths? 7 sixths? 8 sixths?

21. What is 1 seventh of 21?

22. If 1 seventh of 21 is 3, what is 2 sevenths of 21? 3 sevenths? 4 sevenths? 5 sevenths? 6 sevenths?

23. If you can buy 1 pound of candies for 12 cents, what part of a pound can you buy for 1 cent? for 2 cents? for 3 cents? for 5 cents? for 8 cents? for 10 cents?

24. If a coat cost $20, and a pair of pantaloons 1 fourth as much, what will the pantaloons cost?

25. If 7 barrels of cider cost $28, what part of $28 will 1 barrel cost? 4 barrels? 7 barrels? 5 barrels?

26. What is 1 seventh of $28? 2 sevenths of 28? 4 sevenths? 5 sevenths? 7 sevenths? 6 sevenths?

27. If 1 pound of cheese cost 6 cents, what will 1 third of a pound cost? 2 thirds?

28. If 12 lemons cost 36 cents, what part of 36 cents will 1 lemon cost? 2 lemons? 4 lemons? 5 lemons? 8 lemons? 10 lemons? 9 lemons? 7 lemons? 14 lemons?

29. What is 1 twelfth of 36? 2 twelfths of 36? 4 twelfths? 5 twelfths? 6 twelfths? 9 twelfths? 10 twelfths? 14 twelfths?

30. What do you understand by 1 third? 2 thirds?

ANSWER.—When a thing has been divided into three equal parts, 1 of these parts is called 1 *third*, and 2 of these parts are called 2 *thirds*.

31. What do you understand by 1 half?

32. What do you understand by 1 fourth? 2 fourths? 3 fourths?

33. What do you understand by 1 fifth? 2 fifths? 3 fifths? 4 fifths?

34. How many thirds make a whole 1?

35. How many fourths in 1?

36. What do you understand by 2 sixths ? 4 sixths ? 5 sixths ?

37. What do you understand by 3 sevenths ? 2 sevenths ? 4 sevenths ? 5 sevenths ?

38. How many sixths in 1 ?

39. How many ninths in 1 ?

40. How many eighths in 1 ?

41. How many sevenths in 1 ?

42. How many tenths in 1 ?

43. How many twentieths in 1 ?

44. What do you understand by 7 twelfths ? 6 twelfths ? 9 twelfths ? 8 twelfths ?

45. James had 9 marbles, and Jacob had 2 thirds as many ; how many had he ?

46. Mary bought 12 candies, and Sarah bought 2 thirds as many ; how many did Sarah buy ?

47. Rachel has 12 primers, Mary 3 fourths as many, and Anthony 2 thirds as many as Mary ; how many have Mary and Anthony respectively ?

48. Albert is 15 years old, and Ebenezer is 4 fifths as old ; how old is he ?

49. Augustus has 40 cents, and Augusta has 5 eighths as many ; how many has she ?

50. Morgan had 36 marbles, and gave 4 sixths of them to Martin ; how many did he give to Martin, and how many had he left ?

51. Moses has 24 fire-crackers, and Nathan has 7 sixths as many ; how many has he ?

52. Mifflin had 45 cents, and Matthew had 5 ninths as many ; how many had he ?

53. Dubois is 20 years old, and his father is 9 fifths as old ; what is his father's age ?

54. A farmer had 84 sheep, and a wolf killed 1 third of them ; how many had he remaining ?

55. In a certain school there are 12 girls, and 7 fourths as many boys ;—required the number of boys, and the number of boys and girls together.

56. In a certain recitation 36 questions were asked, and 1 ninth of them answered wrong; how many were correctly answered?

57. 4 fifths of all the words given out in a spelling lesson were spelled correctly, and 8 were misspelled; of how many words did the lesson consist?

58. Montgomery bought 9 filberts for 1 cent; what part of a cent did 1 cost? 2? 3? 6? 7? 9?

59. A horse was bought for $60, and sold for 7 fifths of what it cost; how much was the gain?

60. A man received 140 dollars for 14 weeks' labor, and paid 1 fifth of it for board; how much did he save a week?

61. 9 is 1 fourth of what number?

62. How many are 7 eighths of 24?

63. Mr. A's wife is 40 years old, and 9 eighths of her age equals his; what is his age?

64. What is 2 ninths of 36? 4 ninths? 3 fourths? 4 sixths? 5 sixths? 4 twelfths? 9 twelfths?

65. 3 fourths of 24 are how many times 3?

66. How many are 3 fourths of 48? 4 sixths? 5 eighths? 7 eighths? 6 eighths? 5 sixths? 2 thirds?

67. 3 ninths of 27 are how many? 4 ninths? 7 ninths? 8 ninths? 2 thirds?

68. 5 sevenths of 63 are how many times 3?

69. 3 eighths of 64 are how many times 6?

70. 9 thirds of 18 are how many times 3?

71. 4 fifths of 25 are how many times 2?

72. 6 ninths of 18 are how many times 6?

73. 7 ninths of 90 are how many times 2?

74. 4 thirds of 39 are how many times 2?

75. 15 seventeenths of 34 are how many times 6

76. How many times 17 are 17 eighteenths of 36?

77. How many times 8 are 12 thirteenths of 26?

78. How many times 5 are 10 thirds of 36?

79. How many times 4 are 2 thirds of 27, — 2?

80. How many times 6 are 3 halves of 48, + 12?

81. Stephen having 40 apples, gave 3 fifths of them to one companion, and 3 eighths of them to another; how many had he remaining?

82. A had $120; 1 third of it he spent for a watch, 1 fourth of it for a suit of clothes, and 3 tenths of it for a sleigh; how much had he remaining?

83. Mr. B, being asked the cost of his hat, replied, 2 thirds of $30 dollars is 4 times its cost; required the cost of the hat.

84. 14 ninths of $27 is equal to 7 times the cost of a pair of boots; required the cost of the boots.

85. An individual, having $90 on interest, received 2 forty-fifths of the principal for the interest; how much interest did he receive?

86. The interest received on $360, was 1 eighteenth of the principal; what was the interest?

87. B is worth $2000, and 3 fourths of his fortune is 3 times A's; required A's fortune.

88. 3 eighths of the number of hours in a day, is 3 times the number of hours I work; how many hours do I work?

89. A pole, whose length is 16 feet, is in the air and water; and 3 fourths of the whole length — 4 feet, equals what is in the air; required the length in the water.

90. 3 fifths of $2000, + $120, equals B's fortune; how much is B worth?

91. The building of a certain house cost $560, and 4 sevenths of this, + $80 is 1 tenth of the cost of the farm on which it stands. What was the cost of the farm?

92. 5 eighths of 72, + 13, are how many times 2?

93. The interest received on $960 dollars for 5 years, was equal to 1 third of the principal; what was the interest yearly received?

94. What will 2 thirds of 12 pounds of coffee cost, at 13 cents a pound?

95. What will 3 fourths of a gallon of alcohol cost, at 9 cents a pint?

96. What will 1 sixteenth of a bushel of flax-seed cost, at 5 cents a pint?

97. How much will 7 fifteenths of 30 pine-apples cost, at 2 dimes each?

98. How much will 7 ninths of a hogshead of molasses cost, at 4 dimes a gallon?

99. What will 3 fifths of 100 oranges cost, at 1 half dime or (5 cents) each?

100. If 1 pennyweight of gold cost $2, what will 2 fifths of an ounce cost?

101. What will be the cost of 2 thirds of 36 pounds of butter, at 2 dimes a pound?

102. 2 thirds of 24, + 3 fourths of 16, are how many times 7?

103. 2 thirds of 30, + 3 fifths of 40, are how many times 8?

104. 3 sevenths of 21, + 3 eighths of 40, are how many times 6?

105. How far can I walk in 3 eighths of a day, at the rate of 3 miles an hour?

106. If Marcus earn 1 dime in an hour, how much will he earn in 3 eighths of a day?

107. If a horse travel 10 miles in an hour, how many times 10 miles can he travel in 5 twelfths of a day?

108. How many cents will 1 quart of gin cost, if 1 gill cost 15 mills?

109. How many dollars will 4 sixths of a pound of gold cost, if 1 pennyweight cost 12 dimes?

110. How many eagles will 25 fourths of a gallon of brandy cost, at 1 half dime a gill?

LESSON II.

1. If 1 third of an orange cost 2 cents, what will 1 orange cost?

SOLUTION.—If 1 third of an orange cost 2 cents; 3 thirds, which is a whole orange, will cost 3 times 2 cents, which are 6 cents.

2. If 1 half of a pound of raisins cost 8 cents, what will 1 pound cost?

3. Bought 1 third of a barrel of sugar for $3; what will 2 thirds of a barrel cost at the same rate?

4. If 1 third of a pound of pork cost 5 cents, what will 2 pounds cost?

5. 2 is 1 third of what number?

6. 5 is 1 half of what number?

7. If 1 fourth of a lemon cost 2 cents, what will 1 cost?

8. If 1 fourth of a melon cost 5 cents, what will 1 cost?

9. 3 is 1 fourth of what number?

10. 7 is 1 third of what number?

11. 12 is 1 fifth of what number?

12. 7 is 1 fourth of what number?

13. What will 4 fifths of a pound of cinnamon cost, if 1 fifth of a pound cost 5 cents?

14. If 1 fifth of a yard of cloth cost $2, what will a yard cost?

15. If 1 sixth of a gallon of vinegar cost 2 cents, what will a gallon cost?

16. A man, being asked the value of his horse, said that 1 eighth of its value was $12; what was the value of the horse?

17. A man gave 15 cents for his lodging, which was 1 seventh as much as his breakfast cost him; what did he give for his breakfast?

18. Bought 1 eighth of a yard of cloth for 4 dimes; what will a yard cost at that rate?

19. If 1 tenth of a yard of cloth cost 47 cents, how much is that a yard?

20. What will 1 yard of cloth cost, if 1 ninth of a yard cost 5 cents?

21. What will 1 bushel of corn cost, if 1 seventh of a bushel cost 5 cents?

22. What will a hogshead of molasses cost, if 1 eighth of a hogshead cost $3?

23. What will be the cost of 2 cords of wood, if 1 eleventh of a cord cost 30 cents?

24. If 1 twelfth of the distance from Albany to Wilbraham is 9 miles, what is that distance?

25. 9 is 1 tenth of what number?

26. 15 is 1 seventh of what number?

27. 16 is 1 fifth of what number?

28. 12 is 1 fifth of 6 times what number?

29. 15 is 1 sixth of 5 times what number?

30. 18 is 1 fourth of 6 times what number?

31. 10 is 1 eighth of 20 times what number?

32. 15 is 1 seventh of 5 times what number?

33. 20 is 1 eighth of 16 times what number?

34. 30 is 1 third of 6 times what number?

35. A boy's hat cost $3, which was 1 fifth of the cost of his coat. The cost of the coat is required.

36. Mr. B's saddle cost $9, which was 1 fortieth of 6 times the cost of his horse. The cost of the horse is required.

37. Henry gave 5 cents for a piece of pie, which was 1 twentieth of 4 times as much as his breakfast cost him; what was the cost of his breakfast?

38. A man, being asked his age, answered, that his youngest son's age, was 12 years which was just 1 twelfth of 3 times his age. Required the father's age.

39. Mrs. B's shawl cost $9, which was 1 tenth of 3 times the cost of her dress; what was the cost of her dress?

40. John says to James, who is now 10 years old, your age is 1 eighth of 4 times my age. How old is John?

LESSON III.

1. If 2 thirds of a melon cost 4 cents. what will 1 melon cost?

SOLUTION.—If 2 thirds of a melon cost 4 cents, 1 third will cost 1 half of 4 cents, which is 2 cents. If 1 third of a melon cost 2 cents, 3 thirds which is a whole one, will cost 3 times 2 cents, which are 6 cents.

2. If 2 thirds of an orange cost 6 cents, what will 1 third of an orange cost?

3. If 3 fourths of a pound of candies cost 9 cents, what will 1 fourth of a pound cost?

4. If 4 thirds of a pound of spice cost 16 cents, what will 1 third of a pound cost?

5. If 3 fourths of a pound of cinnamon cost 12 cents, what will 1 fourth of a pound cost?

6. If $4 will buy 2 fifths of a barrel of fish, what will 1 fifth of a barrel cost?

7. What will 1 sixth of a yard of cloth cost, if 4 sixths of a yard cost 120 cents?

8. What will 1 seventh of a hogshead of molasses cost, if 5 sevenths of a hogshead cost $15?

9. What will 1 ninth of a pound of coffee cost, if 8 ninths of a pound cost 12 cents?

10. If 2 thirds of a barrel of fish cost $8, what will 1 third of a barrel cost?

11. If 1 third of a barrel of flour cost $3, what will 1 barrel cost?

12. If 2 thirds of a barrel of fish cost $8, what will 1 barrel cost?

13. If 3 fourths of a bushel of wheat cost 9 dimes, what will 1 bushel cost?

14. If 4 fifths of a box of raisins cost 12 dimes, what will 1 box cost?

15. If 6 eighths of a yard of broadcloth cost 30 dimes, how many dollars will 1 yard cost?

16. 8 is 2 thirds of what number?

SOLUTION.—If 8 is 2 thirds of some number, 1 third of that number is 1 half of 8, which is 4; and, if 1 third of that number is 4, 3 thirds, which is that number, is 3 times 4, which are 12. Therefore, 8 is 2 thirds of 12.

17. 10 is 2 sevenths of what number?
18. 9 is 3 fourths of what number?
19. 12 is 3 fourths of what number?
20. 12 is 6 elevenths of what number?
21. 14 is 7 eighths of what number?
22. 14 is 2 sevenths of what number?
23. 6 is 3 tenths of what number?
24. 9 is 3 sevenths of what number?
25. 15 is 5 sixths of what number?
26. 15 is 3 halves of what number?
27. 18 is 9 eighths of what number?
28. 20 is 5 thirteenths of what number?
29. 24 is 8 fifths of what number?
30. 26 is 13 ninths of what number?
31. 2 thirds of 12 is 2 fifths of what number?
32. 3 fourths of 12 is 3 eighths of what number?
33. 3 fourths of 8 is 2 sevenths of what number?
34. 3 fifths of 25 is 5 fourths of what number?
35. 2 sevenths of 14 is 4 ninths of what number?
36. 4 sevenths of 21 is 4 fifteenths of what number?
37. 2 thirds of 15 is 5 fourths of what number?
38. 7 eighths of 48 is 3 halves of what number?
39. 8 ninths of 36 is 4 fifths of what number?
40. 7 thirds of 18 is 3 fifths of what number?
41. A watch cost $16, and 3 halves of the cost of the watch is 8 thirds of the cost of the chain. What was the cost of the chain?
42. A horse was sold for $96, which was 8 sevenths of what it cost; what was the cost of the horse?
43. In a certain school there are 36 ladies, and 5 fourths of the number of ladies equals 3 fifths of the number of gentlemen. How many gentlemen were there in the school?
44. Mary is 14 years old, and 9 sevenths of her age is 2 thirds of Margaret's age; what was her age?
45. A piece of cloth containing 12 yards was sold

for $60, which was 5 fourths of what it cost; what did it cost, and what was the gain on each yard?

46. A has 48 geese, and 3 fourths of his number is equal to 9 sevenths of B's number; how many geese has B?

47. The head of a fish is 12 inches long, and 3 fourths of the length of the head is 3 fifteenths of the length of the body. Required the length of the fish.

48. A farm was sold for $1200, which was only 6 sevenths of what it was worth. How much was lost by the bargain?

49. $48 is 3 fifths of the cost of 12 yards of cloth; how must it be sold a yard to gain $16 on the whole?

50. A man gave $60 for a suit of clothes, which was 1 fifth of his yearly income; 1 sixth of the remainder he spent for a watch, and what then remained was 4 fifths of his brother's yearly income. What was the yearly income of each?

LESSON IV.

2 thirds	is written thus			$\frac{2}{3}$.
1 half	"	"	"	$\frac{1}{2}$.
1 third	"	"	"	$\frac{1}{3}$.
1 fourth	"	"	"	$\frac{1}{4}$.
1 sixth	"	"	"	$\frac{1}{6}$.
1 seventh	"	"	"	$\frac{1}{7}$.
3 fourths	"	"	"	$\frac{3}{4}$.
7 eighths	"	"	"	$\frac{7}{8}$.
9 tenths	"	"	"	$\frac{9}{10}$.
5 sevenths	"	"	"	$\frac{5}{7}$.
2 fifths	"	"	"	$\frac{2}{5}$.
5 thirds	"	"	"	$\frac{5}{3}$.
&c.	"	"	"	&c.

Remark.—The above expressions are called *fractions*. The figure above the short horizontal line is called the *numerator*, and the figure below the line is called the *denominator*. For example, in the fraction $\frac{3}{4}$; the 3 is the numerator, and the 4 is the denominator.

The *denominator* of a fraction shows into how many parts the thing is divided; and the *numerator* shows how many of these parts are taken.

1. If you cut an orange into 3 equal parts, what 1 of these parts called?

2. If a lemon be cut into 4 equal pieces, what will 1 of these pieces be called? 2? 3? 4?

3. If a bushel of apples be divided into 6 equal parts, what will 1 of these parts be called? 3? 4? 6?

4. If a basket of peaches be divided into 8 equal parts, what will 3 of these parts be called? 5? 6? 7?

5. How can you find 2 thirds of an apple?

6. How can you find 3 fourths of an orange?

7. In $\frac{4}{2}$ how many times 1?

Solution.—In 1 there are 2 halves, therefore 1 half of the number of halves equals the number of ones. 1 half of 4 is 2; therefore, $\frac{4}{2}$ equals 2

8. In $\frac{6}{2}$ how many times 1?
9. In $\frac{8}{2}$ how many times 1?
10. In $\frac{10}{2}$ how many times 1?
11. In $\frac{12}{2}$ how many times 1?
12. In $\frac{14}{2}$ how many times 1?
13. In $\frac{6}{3}$ how many times 1?
14. In $\frac{12}{3}$ how many times 1?
15. In $\frac{9}{3}$ how many times 1?
16. In $\frac{15}{3}$ how many times 1?
17. In $\frac{18}{2}$ how many times 1?
18. In $\frac{18}{3}$ how many times 1?
19. In $\frac{21}{3}$ how many times 1?
20. In $\frac{8}{4}$ how many times 1?
21. In $\frac{12}{4}$ how many times 1?
22. In $\frac{16}{4}$ how many times 1?
23. In $\frac{20}{4}$ how many times 1?
24. In $\frac{24}{4}$ how many times 1?

25. In $\frac{36}{4}$ how many times 1?
26. In $\frac{5}{5}$ how many times 1?
27. In $\frac{10}{5}$ how many times 1?
28. In $\frac{25}{5}$ how many times 1?
29. In $\frac{15}{5}$ how many times 1?
30. In $\frac{55}{5}$ how many times 1?
31. In $\frac{60}{5}$ how many times 1?
32. In $\frac{18}{6}$ how many times 1?
33. In $\frac{36}{9}$ how many times 1?
34. In $\frac{48}{8}$ how many times 1?
35. In $\frac{50}{10}$ how many times 1?
36. In $\frac{45}{15}$ how many times 1?
37. In $\frac{72}{18}$ how many times 1?
38. In $\frac{100}{2}$ how many times 1?
39. In $\frac{56}{7}$ how many times 1?
40. In $\frac{63}{9}$ how many times 1?
41. In $\frac{7}{2}$ how many times 1?
42. In $\frac{9}{2}$ how many times 1?
43. In $\frac{15}{4}$ how many times 1?
44. In $\frac{19}{3}$ how many times 1?
45. In $\frac{23}{4}$ how many times 1?
46. In $\frac{21}{5}$ how many times 1?

REMARK.—This is called reducing fractions to *whole*, or *mixed numbers*. A mixed number is a whole number with a fraction added to it. Thus $8\frac{1}{2}$ is a *mixed number*.

Whenever the *numerator* is less than the denominator, the value is less than a unit, and the expression is called a *proper fraction*;—but when the *numerator* is greater than the denominator the value is greater than a unit, and the expression is called an *improper fraction*.

47. What kind of a fraction is it called, when the *numerator* is less than the denominator?

48. What kind of a fraction is it called, when the *denominator* is *less* than the numerator?

49. When is the value of a fraction *greater* than a unit?

50. When the *denominator* is *greater* than the numerator, what kind of a fraction is it called?

51. What kind of a fraction is it called, when the *numerator* is *larger* than the denominator?

52. Reduce $\frac{57}{4}$ to a mixed number.

53. Reduce $\frac{89}{12}$ to a mixed number.

54. Reduce $\frac{94}{8}$ to a mixed number.

55. Reduce $\frac{25}{3}$ to a mixed number.

56. Reduce $\frac{37}{9}$ to a mixed number.

57. Reduce $\frac{47}{4}$ to a mixed number.

58. Reduce $\frac{78}{7}$ to a mixed number.

59. Reduce $\frac{37}{6}$ to a mixed number.

60. Reduce $\frac{34}{8}$ to a mixed number.

LESSON V.

1. James had $\frac{3}{4}$ of an apple, and John gave him $\frac{3}{4}$ more; how many had he then?

2. Mary had $\frac{5}{7}$ of an orange, and her father gave her $\frac{2}{7}$ of an orange more; how many had she then?

3. Robert had $\frac{2}{3}$ of a melon, and bought $\frac{2}{3}$ of another one; how much had he then?

4. Susan had $\frac{5}{7}$ of a pint of walnuts, and gave $\frac{3}{7}$ of a pint to her sister; how much had she left?

5. James bought $\frac{19}{4}$ of a pound of candies, and on his way home ate $\frac{3}{4}$ of a pound; how many had he left?

6. John gave [illegible] of a pound of raisins to James, $\frac{4}{5}$ of a pound to Mary, and kept $\frac{3}{5}$ of a pound himself; how many had he at first?

7. Mortimer gave $\frac{4}{5}$ of a dollar for a hat, \$1$\frac{2}{5}$ for a vest, and had \$3$\frac{4}{5}$ remaining; how much had he at first?

8. Jane had 5 pounds of cinnamon, and Harriet had 2$\frac{3}{4}$ pounds; how many more had Jane than Harriet?

9. Henry gave $\frac{2}{5}$ of a dollar for his breakfast, $\frac{3}{5}$ of a dollar for his dinner, and $\frac{4}{5}$ of a dollar for his supper; how much did his day's board cost him?

10. $\frac{8}{9} + \frac{4}{9}$ are how many?

11. $\frac{7}{8} + \frac{6}{8}$ are how many?

12. $\frac{5}{8} + \frac{6}{8}$ are how many?

13. $\frac{9}{10} + \frac{8}{10}$ are how many?

14. $\frac{3}{4} + \frac{2}{4}$ are how many?

15. $\frac{16}{3} + \frac{4}{3}$ are how many?

16. $\frac{14}{3} + \frac{5}{3}$ are how many?

17. $\frac{14}{8} + \frac{16}{8}$ are how many?

18. $\frac{9}{4} + \frac{13}{4}$ are how many?

19. $\frac{7}{8} + \frac{6}{8} + \frac{5}{8}$ are how many?

20. $\frac{3}{8} + \frac{6}{8} + \frac{4}{8}$ are how many?

21. $\frac{7}{10} + \frac{8}{10} + \frac{9}{10}$ are how many?

22. $\frac{3}{7} + \frac{4}{7} + \frac{6}{7}$ are how many?

23. $\frac{8}{9}$ less $\frac{4}{9}$ are how many ninths

24. $\frac{27}{8}$ less $\frac{17}{8}$ are how many?

25. $\frac{19}{4} - \frac{3}{4}$ are how many?.

26. $\frac{27}{3} - \frac{2}{3}$ are how many?

27. $\frac{37}{5} - \frac{3}{5}$ are how many?

28. $\frac{14}{3} - \frac{2}{3}$ are how many?

29. $\frac{37}{8} - \frac{5}{8}$ are how many?

30. $\frac{19}{7} - \frac{2}{7}$ are how many?

31. $\frac{46}{9} - \frac{10}{9}$ are how many?

32. $\frac{47}{8} - \frac{6}{8}$ are how many?

33. $\frac{57}{12} - \frac{37}{12}$ are how many?

34. $\frac{9}{4} + \frac{7}{4} - \frac{3}{4}$ are how many?

35. $\frac{7}{8} + \frac{5}{8} - \frac{2}{8}$ are how many?

36. $\frac{4}{5} + \frac{17}{5} - \frac{9}{5}$ are how many?

37. $\frac{14}{15} - \frac{10}{15} + \frac{12}{15}$ are how many?

38. $\frac{2}{5}$ of 60 — $\frac{3}{4}$ of 24 are how many?

39. $\frac{7}{8}$ of 40 — $\frac{2}{5}$ of 10 are how many?

40. $\frac{4}{5}$ of 15 + $\frac{2}{3}$ of 9 — $\frac{3}{4}$ of 12 are how many

LESSON VI.

1. At $\frac{2}{3}$ of a cent a piece, what will 2 apples cost?

SOLUTION.—If 1 apple cost $\frac{2}{3}$ of a cent, 2 apples will cost twice $\frac{2}{3}$ of a cent, which are $\frac{4}{3}$ or $1\frac{1}{3}$ cents.

2. At $\frac{2}{3}$ of a cent a piece, what will 5 apples cost?

3. At $\frac{4}{5}$ of a dime a pound, what will 10 pounds of candies cost?

4. At $1\frac{3}{4}$ dimes a pound, what will 8 pounds of starch cost?

5. At $\frac{2}{5}$ of a cent a piece, what will 25 filberts cost?

6. At $\frac{8}{9}$ of a dime a piece, what will 8 chickens cost?

7. At $\frac{2}{5}$ of a dollar a yard, what will 15 yards of linen cost?

8. If a man spend $\frac{3}{4}$ of a dollar a day, how much, at this rate, will he spend in 23 days?

9. If a man receive $\frac{2}{3}$ of an eagle in a week, how much will he receive in 52 weeks?

10. If 1 pound of sugar cost $1\frac{1}{3}$ dimes, what will 12 pounds cost?

11. At $5\frac{2}{3}$ cents a pound, what will 6 pounds of beef cost?

12. At $9\frac{3}{4}$ cents a pound, what will 8 pounds of pork cost?

13. At $6\frac{2}{5}$ cents each, what will 12 lemons cost?

14. At $7\frac{3}{5}$ cents each, what will 20 rabbits cost?

15. At $12\frac{1}{2}$ cents a dozen, what will 4 dozen eggs cost?

16. At $11\frac{3}{8}$ cents a pound, what will 6 pounds of honey cost?

17. At $\$7\frac{4}{5}$ a barrel, what will 10 barrels of tobacco cost?

18. At $\$9\frac{3}{4}$ a barrel, what will 10 barrels of pork cost?

19. What will 6 boxes of raisins cost, at $3[illegible] a box?

20. What will 14 bushels of wheat cost, at \$1$\frac{2}{3}$ a bushel?
21. What cost 7 barrels of cider, at \$3$\frac{3}{4}$ a barrel?
22. If a barrel of flour cost \$4, what will 5$\frac{3}{4}$ barrels cost?
23. 5 times 4 and $\frac{2}{4}$ of 4 are how many?
24. 7 times 6 and $\frac{4}{6}$ of 6 are how many?
25. 9 times 7 and $\frac{5}{7}$ of 7 are how many?
26. 12 times 9 and $\frac{8}{9}$ of 9 are how many?
27. 5 times 10 and $\frac{4}{5}$ of 10 are how many?
28. 13 times 4 and $\frac{3}{4}$ of 4 are how many?
29. 8 times 7 and $\frac{3}{7}$ of 7 are how many?
30. 10 times 13 and $\frac{11}{13}$ of 13 are how many?
31. 7 times 20 and $\frac{3}{5}$ of 20 are how many?
32. How many are 4 times $\frac{2}{3}$?
33. How many are 4 times 2$\frac{2}{3}$?
34. How many are 3 times 4$\frac{3}{4}$?
35. How many are 5 times 3$\frac{2}{7}$?
36. How many are 7 times 9$\frac{2}{3}$?
37. How many are 8 times 12$\frac{5}{7}$?
38. How many are 9 times 10$\frac{3}{4}$?
39. How many are 6 times 12$\frac{2}{3}$?
40. How many are 12 times 9$\frac{7}{8}$?

LESSON VII.

1. If you give to 6 persons, each $\frac{2}{3}$ of a dollar, how many dollars will it take?
2. What will be the cost of 4 yards of cloth, at $\frac{3}{4}$ of a dollar a yard?
3. If 1 yard of cloth cost \$1$\frac{2}{5}$, what will 10 yards cost?
4. How many oranges will it require to give to 9 boys, if to one boy you give 1$\frac{1}{3}$ oranges?
5. How many barrels of flour does that man give away, who gives to each of 12 beggars $\frac{3}{8}$ of a barrel?

6. Anthony gave each of his 7 companions $\frac{2}{3}$ of a pound of candies, and had $\frac{1}{3}$ of a pound left; how many pounds had he at first?

7. Thornton gave to each of 9 beggars $\frac{5}{6}$ of a dollar, and had $7 remaining; how much had he at first?

8. James gave $\frac{1}{12}$ of an orange to Jackson, $\frac{5}{12}$ to Joseph, and $\frac{3}{12}$ to John; what part of an orange had he remaining?

9. Harmon meeting 3 poor women and 5 poor men, gave to each woman $\frac{3}{5}$ of a dollar, and to each man $\frac{4}{5}$ of a dollar, and then had only $4 remaining; how much had he at first?

10. How many quarts of chestnuts must that boy have, who gives to each of 20 persons $\frac{7}{8}$ of a quart, and has 7 quarts remaining?

11. Mary gave to each of her 12 companions as many pinks as she then had roses, which were 2, and had no flowers remaining but her roses. How many flowers had she at first?

12. What cost 1 quart of vinegar, if 1 pint cost $\frac{2}{3}$ of a cent?

13. If 1 gill of molasses cost $\frac{3}{5}$ of a cent, what will 2 quarts cost?

14. If 2 pints of beans cost 4 cents, what will 1 peck cost?

15. If 3 pecks of buckwheat cost 96 cents, what will 1 pint cost?

16. What will 10 yards of silesia cost, if 1 yard cost $18\frac{2}{5}$ cents?

17. What will $4\frac{2}{3}$ yards of silk cost, if 1 yard cost dimes?

18. What will $\frac{3}{5}$ of a yard of muslin cost, if 1 yard cost 10 cents?

19. What will 7 spools of thread cost, if 1 spool cost $7\frac{5}{8}$ cents?

20. What will $8\frac{2}{3}$ yards of silk cord cost, at 6 cents a yard?

21. If 1 yard of wadding cost 5 cents, what will $9\frac{4}{5}$ yards cost?

22. What will $6\frac{3}{4}$ yards of muslin cost, if 1 yard cost 8 cents?

23. What will $8\frac{3}{5}$ pounds of veal cost, at 5 cents a pound?

24. How much will $9\frac{3}{4}$ barrels of cider cost, at $4 a barrel?

25. What would be the cost of 12 window-sash, at $$3\frac{3}{4}$ each?

26. What will $5\frac{3}{4}$ yards of gingham cost, at 4 dimes a yard?

27. What will be the cost of 13 yards of bishop lawn, at $$1\frac{4}{13}$ a yard?

28. What will be the cost of 8 looking glasses, at $$15\frac{3}{4}$ a piece?

29. What amount of money will be required to purchase 30 pounds of rice, at $6\frac{2}{5}$ cents a pound?

30. What will be the cost of 23 pounds of crackers, at $8\frac{2}{4}$ cents a pound?

31. What will 9 barrels of fish cost, at $$12\frac{2}{3}$ a barrel?

32. If 1 grain of gold cost $9\frac{1}{2}$ dimes, what will 1 pennyweight cost?

33. If 1 gold pen cost $$2\frac{2}{3}$, what will 6 cost?

34. How many pounds of meat, at 5 cents a pound can you buy for $$3\frac{4}{5}$?

35. What will be the cost of 3 quarts of nuts, at 64 cents a peck?

36. If a coachman charge $5\frac{1}{2}$ cents a mile, how much must that man pay who rides 12 miles?

37. How many dollars, dimes and cents will 12 yards of cloth cost, at 62 cents a yard?

38. How many dollars and cents will 4 pecks of grass seed cost, if 1 pint cost 5 cents?

39. How much will 13 yards of shalloon cost, at $13\frac{3}{4}$ cents a yard?

40. What will be the cost of 16 bushels of potatoes, at $2\frac{5}{7}$ dimes a bushel?

LESSON VIII.

☞ REMARK.—A fraction may be multiplied, by *multiplying* the *numerator*, (as you have already observed,) or by *dividing* the *denominator*.

1. How many are 5 times $\frac{4}{10}$?

SOLUTION.—5 times $\frac{1}{10}$ is $\frac{1}{2}$; and, if 5 times $\frac{1}{10}$ is $\frac{1}{2}$, 5 times $\frac{4}{10}$ are 4 times $\frac{1}{2}$, which are $\frac{4}{2}$ or 2. Therefore 5 times $\frac{4}{10}$ are 2.

REMARK.—After the pupil has become familiar with this method of solution the following may be adopted.

SOLUTION.—5 times $\frac{4}{10}$ are $\frac{4}{2}$ or 2.

2. How many are 3 times $\frac{13}{9}$?
3. How many are 9 times $\frac{6}{27}$?
4. How many are 5 times $\frac{27}{15}$?
5. How many are 6 times $\frac{27}{12}$?
6. How many are 9 times $\frac{4}{18}$?
7. How many are 9 times $\frac{16}{36}$?
8. How many are 7 times $\frac{39}{21}$?
9. 8 times $\frac{13}{16}$ are how many?
10. 11 times $\frac{13}{22}$ are how many?
11. 13 times $\frac{15}{26}$ are how many?
12. 2 times $\frac{9}{4}$ are how many?
13. 5 times $\frac{17}{20}$ are how many?
14. 6 times $\frac{49}{12}$ are how many?
15. 7 times $\frac{37}{14}$ are how many?
16. 12 times $\frac{13}{36}$ are how many?
17. How many times 5 are 8 times $\frac{40}{16}$?
18. How many times 12 are 9 times $\frac{48}{18}$?
19. How many times 8 are 11 times $\frac{64}{22}$?
20. How many times 100 are 25 times $\frac{400}{50}$?
21. How many times 20 are 35 times $\frac{800}{70}$?
22. 5 times $\frac{48}{5}$ is 4 times Mary's age; what is her age?

23. 13 times $\frac{150}{39}$ equals $\frac{1}{2}$ of the number of dollars a certain wagon cost. Required the cost of the wagon.

24. 25 times $\frac{100}{25}$ equals $\frac{1}{700}$ of the number of men that Gen. Santa Anna had at the battle of Buena Vista. How many men had he?

25. 6 times $\frac{200}{12}$ is $\frac{1}{15}$ of the number of men he had wounded. How many men were wounded?

26. 7 times $\frac{569}{7}$ is $\frac{1}{2}$ of the number of men he had killed. How many were killed?

27. 4 times $\frac{450}{9}$ is $\frac{1}{20}$ of the number of men that Gen. Taylor had. How many had he?

28. 9 times $\frac{67}{9}$ is $\frac{1}{4}$ of the number of men he had killed. How many were killed?

29. 8 times $\frac{15}{16}$ is $\frac{1}{2}$ of what number?

30. 4 times $\frac{25}{12}$ is $\frac{1}{3}$ of what number?

31. A laborer worked 12 months, at the rate of $\$10\frac{3}{4}$ a month; how much did his year's wages amount to?

32. If 2 quarts of wine cost 48 cents, what will 1 gill cost?

33. How much ought I to pay for 3 oranges, at $\frac{3}{4}$ of a cent a piece?

34. If a certain piece of work can be performed in 96 hours, how many days will be required to perform it by working 6 hours a day?

35. If 1 man can dig a ditch in 15 days, how long will it take 5 men to do it?

36. If a certain quantity of provision serve a family of 4 persons 16 days, how long would it last a family of 8 persons?

37. If 8 men can perform a certain piece of work in 56 days, in how many days can 112 men do the same?

38. If 3 men can plow 18 acres in 6 days, in how many days could 9 men do the same?

39. 4 men can mow a certain field in $6\frac{1}{4}$ days, in how many days can 5 men perform the same work?

40. A man bought 6 barrels of cider, at \$$3\frac{2}{3}$ a barrel; how many boxes of butter, at \$4 a box, will it take to pay for it?

41. A merchant bought 6 yards of cloth and sold it for \$20 which was $\frac{10}{9}$ of what it cost; what did it cost a yard?

42. Bought 36 yards of cloth, and sold $\frac{5}{9}$ of it for \$25, which was $\frac{5}{4}$ of what it cost; how much would I have gained by selling the whole at the same rate?

43. 7 men in $\frac{5}{7}$ of a day can earn \$10, how long would it take 1 man to earn the same?

44. James is $3\frac{3}{5}$ years of age, which is $\frac{1}{5}$ of the age of Henry; and Henry is 9 times as old as George. What is the age of Henry and George respectively?

45. $\frac{1}{3}$ of 36 is 3 times $\frac{1}{2}$ of what number?

46. $\frac{1}{2}$ of 32 is $\frac{2}{3}$ of 3 times what number?

47. $\frac{2}{3}$ of 60 is $\frac{2}{5}$ of twice what number?

48. $\frac{3}{4}$ of 40 is $\frac{3}{7}$ of as many dollars as Mr. B's horse cost; what was the cost of his horse?

49. A person, being asked his age, said, that $\frac{3}{4}$ of 80 was $\frac{2}{3}$ of ten times his age. What was his age?

50. Morgan is 20 years old, and $\frac{4}{5}$ of his age is $\frac{4}{7}$ of the age of his brother. What was his brother's age?

LESSON IX.

1. How many thirds are there in 3?

Solution.—In 1 there are 3 thirds, and in 3 there are 3 times 3 thirds, which are $\frac{9}{3}$.

The following solution is preferred to the above.

Solution.—In 1 there are 3 thirds, therefore, 3 times the number of whole ones, equal the number of thirds. 3 times $\frac{3}{3}$ are $\frac{9}{3}$.

2. How many fourths are there in 3?

3. How many halves are there in 6?

4. How many fifths are there in 4 ? in 5 ?
5. How many fifths are there in 7 ? in 8 ?
6. How many sixths are there in 4 ? in 3 ? 5 ?
7. How many sevenths are there in 2 ? in 4 ? in 6 ?
8. How many eighths are there in 7 ? in 4 ? in 5 ?
9. How many fifteenths are there in 2 ? in 3 ? in 6 ?
10. How many tenths are there in 4 ? in 6 ? 7 ?
11. How many fourths are there in 3 and $\frac{2}{4}$?
12. How many thirds are there in 4 and $\frac{1}{3}$?
13. How many th rds are there in 3 and $\frac{2}{3}$?
14. How many halves are there in 8 and $\frac{1}{2}$?
15. Reduce $6\frac{5}{6}$ to an improper fraction.
16. Reduce $9\frac{3}{5}$ to an improper fraction.
17. Reduce $7\frac{3}{7}$ to an improper fraction.
18. Reduce $5\frac{3}{4}$ to an improper fraction.
19. Reduce $4\frac{3}{8}$ to an improper fraction.
20. Among how many individuals, can $5\frac{3}{7}$ bushels of wheat be distributed, providing each receives $\frac{1}{7}$ of a bushel ?
21. Among how many boys, can $7\frac{3}{8}$ oranges be divided, providing each receives $\frac{1}{8}$ of an orange ?
22. 8 and $\frac{2}{9}$ are how many times $\frac{2}{9}$?
23. $9\frac{6}{8}$ are how many times $\frac{2}{8}$
24. $9\frac{3}{5}$ are how many times $\frac{3}{5}$?
25. $7\frac{4}{5}$ are how many times $\frac{3}{5}$?
26. $12\frac{2}{9}$ are how many times $\frac{5}{9}$?
27. $7\frac{5}{7}$ are how many times $\frac{3}{7}$?
28. $7\frac{2}{4}$ are how many times $\frac{3}{4}$?
29. $4\frac{3}{6}$ are how many times $\frac{9}{6}$?
30. $10\frac{4}{5}$ are how many times $\frac{3}{5}$?
31. $8\frac{6}{9}$ are how many times $\frac{3}{9}$?
32. $12\frac{6}{7}$ are how many times $\frac{3}{7}$?
33. 4 times $3\frac{1}{3}$ are how many times $\frac{2}{3}$?
34. 9 times $1\frac{3}{5}$ are how many times $\frac{4}{5}$?
35. 5 times $6\frac{2}{5}$ are how many times $\frac{2}{5}$?
36. 8 times $8\frac{2}{3}$ are how many times $\frac{2}{3}$?
37. 6 times $2\frac{4}{7}$ are how many times $\frac{3}{7}$?

38. A boy distributed $9\frac{3}{5}$ apples equally among his companions, giving to each $\frac{4}{5}$ of an apple ; how many companions had he ?

39. Homer distributed $\$12\frac{4}{5}$ equally among some poor women, giving to each $\$1\frac{3}{5}$; how many women were there ?

40. Mary gave $\frac{3}{5}$ of a pie to each of her 9 visitors ; how many pies did it take ?

41. Bought 8 yards of cloth, at $\$5\frac{2}{3}$ a yard ; how many yards of silk, worth $\$1\frac{1}{3}$ a yard will it take to pay for it ?

42. Bought 9 yards of cloth, worth $\$1\frac{5}{6}$ a yard, and paid for it with raisins, at $\$1\frac{3}{6}$ a box ; how many boxes did it take ?

43. How many bushels of turnips, at $\$\frac{2}{5}$ a bushel can be bought for 8 bushels of apples, at $\$\frac{3}{5}$ a bushel ?

44. How many apples, at $\frac{3}{4}$ of a cent each, can be bought for 6 oranges, at $2\frac{1}{4}$ cents apiece ?

45. How many yards of cloth, at $\$\frac{5}{7}$ a yard, can be bought for 10 boxes of butter, at $\$5\frac{3}{7}$ a box ?

46. How many geese, at $\$\frac{7}{8}$ each, can be bought for 14 ducks, at $\$\frac{3}{8}$ apiece ?

47. How many boxes of cheese, worth $\$2\frac{3}{7}$ a box, may be had for 17 boxes of butter, at $\$1\frac{3}{7}$ a box:

48. How many barrels of flour, worth $\$5\frac{2}{3}$ a barrel, may be had for 17 bunches of cotton, at $\$1\frac{1}{3}$ a bunch ?

49. How many sheep, at $\$1\frac{1}{3}$ a head, may be had for 8 calves, at $\$3\frac{2}{3}$ each ?

50. How many quarts of alcohol, at $\frac{2}{3}$ of a cent a pint, may be had for 12 quarts of molasses, at $4\frac{1}{3}$ cents a quart ?

LESSON X.

LESSONS NINTH AND TENTH COMBINED.

☞ REMARK.—A fraction may be divided, by *multiplying the denominator*; or by *dividing* the *numerator*.

1. If 2 yards of cloth cost $\$\frac{4}{5}$, what will 1 yard cost?

SOLUTION.—If 2 yards cost $\$\frac{4}{5}$, 1 yard will cost $\frac{1}{2}$ of $\$\frac{4}{5}$, which is $\$\frac{2}{5}$.

2. If 3 apples cost $\frac{6}{7}$ of a cent, what will 1 apple cost?
3. If 2 oranges cost $\frac{6}{7}$ of a cent, what will 1 orange cost?
4. If 3 yards of cord cost $\frac{12}{13}$ of a cent, what will 1 yard cost?
5. If 2 pounds of sugar cost $8\frac{2}{3}$, (or $\frac{26}{3}$ cents,) what will 1 pound cost?
6. If 2 pine-apples cost $14\frac{2}{3}$ cents, what will 1 pine-apple cost?
7. If $\frac{2}{9}$ of a melon is worth 2 oranges, how much is 1 orange worth?
8. If 3 apples are worth $1\frac{3}{7}$ quinces, what is 1 apple worth?
9. How many times 7 are $\frac{21}{23}$?
10. How many times 13 are $7\frac{4}{5}$?
11. How many times 7 are $9\frac{1}{3}$?
12. How many times 21 are $16\frac{4}{5}$?
13. How many times 8 are $33\frac{3}{5}$?
14. How many times 7 are $10\frac{1}{2}$?
15. How many times 11 are $40\frac{1}{3}$?
16. How many times 18 are $14\frac{2}{5}$?
17. How many times 3 are $4\frac{4}{5}$?
18. How many times 9 are $7\frac{5}{7}$?
19. How many times 6 are $6\frac{6}{8}$?
20. How many times 15 are $83\frac{1}{3}$?

21. If 9 oranges are worth $\$\frac{9}{10}$, how many **cents is** 1 orange worth?

22. If 4 boxes of figs cost $\frac{4}{5}$ of an eagle, how many dollars will 1 box cost?

23. If 7 pounds of cheese cost $\$\frac{7}{10}$, how many cents will 1 pound cost?

24. If 3 cakes cost $\frac{9}{10}$ of a dime, what will 1 cake cost?

25. If 4 pounds of chocolate cost $4\frac{4}{5}$ dimes, how many cents is that a pound?

26. What will 1 portfolio cost, if 3 cost $\frac{9}{10}$ of an eagle?

27. If 8 quarts of alcohol cost 32 dimes, how many cents will 2 gills cost?

28. What will 1 pound of sugar cost, if 4 pounds cost $18\frac{2}{3}$ cents?

29. If 6 pounds of cheese cost $31\frac{3}{4}$ cents, what will 1 pound cost?

30. If 12 eggs cost $9\frac{3}{5}$ cents, what will be the cost of 1 egg?

31. If 7 yards of cloth cost $\$24\frac{1}{2}$, what will 1 yard cost?

32. If 5 silk shawls cost $\$27\frac{1}{2}$, how much is that apiece?

33. If 9 pair of boots cost $\$32\frac{2}{5}$, how much is that a pair?

34. If 9 oranges are worth $30\frac{3}{5}$ walnuts, how many walnuts is 1 orange worth?

35. A boy gave 8 apples for $18\frac{2}{3}$ marbles, how many marbles did he get for 1 apple?

36. A boy gave 7 cents for $17\frac{1}{2}$ crackers; how many did he get for 1 cent?

37. Mary gave 10 pins for $23\frac{1}{3}$ chestnuts; how many did she get for 1 pin?

38. If 3 yards of broadcloth are worth $18\frac{6}{7}$ yards of muslin, how many yards of muslin may be had for 1 yard of broadcloth?

39. If John can walk 13 miles while Josiah is walking $32\frac{1}{2}$ miles, how far can Josiah walk while John is walking 1 mile ?

40. If 2 chestnuts are worth $\frac{2}{10}$ of a cent, and 20 chestnuts are worth $\frac{4}{7}$ of a lemon, how many cents is lemon worth ?

41. If 2 oranges cost $\frac{5}{3}$ of a cent, what will 1 orange cost ?

SOLUTION.—If 2 oranges cost $\frac{5}{3}$ of a cent, 1 orange will cost $\frac{1}{2}$ of $\frac{5}{3}$ of a cent, which is $\frac{5}{6}$ of a cent.

42. If 3 yards of linen cost $\$\frac{8}{5}$, what will 1 yard cost ?

43. If 7 yards of tape cost $13\frac{2}{3}$ cents, what will 1 yard cost ?

44. If 2 pints of molasses cost $1\frac{2}{5}$ dimes, how much will 1 gallon cost ?

45. How many times 8 are $6\frac{1}{2}$?

46. How many times 6 are $5\frac{2}{3}$?

47. How many times 4 are $4\frac{1}{2}$?

48. How many times 2 are $13\frac{1}{3}$?

49. How many times 7 are $7\frac{3}{5}$?

50. How many times 8 are 9 times $\frac{27}{18}$?

REMARK.—Multiply as in Lesson Eighth.

51. How many times 3 are 6 times $1\frac{10}{18}$?

52. How many times 9 are 10 times $2\frac{3}{20}$?

53. How many times 7 are 5 times $3\frac{1}{10}$?

54. How many times 5 are 15 times $3\frac{7}{30}$?

55. How many times 7 are 15 times $1\frac{28}{45}$?

56. How many times 5 are 3 times $2\frac{2}{9}$?

57. How many times 6 are 4 times $5\frac{2}{8}$?

58. If $1\frac{2}{3}$ yards of cloth are worth $\$2\frac{1}{2}$, what is 1 yard worth ?

59. If $6\frac{2}{3}$ bunches of grapes are worth 40 cents, how many cents is 1 bunch worth ?

60. If $3\frac{2}{3}$ baskets of peaches are worth $\$5\frac{1}{2}$, what is 1 basket of peaches worth ?

LESSON XI.

1. What is $\frac{1}{3}$ of 2?

SOLUTION.—$\frac{1}{3}$ of 1 is $\frac{1}{3}$; and, if $\frac{1}{3}$ of 1 is $\frac{1}{3}$, $\frac{1}{3}$ of 2 is twice $\frac{1}{3}$, which are $\frac{2}{3}$. Therefore $\frac{1}{3}$ of 2 is $\frac{2}{3}$ of 1.

2. What is $\frac{1}{3}$ of 4? of 8?
3. What is $\frac{1}{4}$ of 2? 3? 5? 7? 9?
4. What is $\frac{1}{2}$ of 3? 5? 7? 9? 11?
5. What is $\frac{1}{5}$ of 2? 3? 4? 7? 8?
6. What is $\frac{1}{6}$ of 2? 3? 5? 7? 9?
7. What is $\frac{1}{7}$ of 2? 3? 5? 4? 6? 9? 11?
8. What is $\frac{1}{8}$ of 2? 4? 3? 5? 6? 7? 9?
9. What is $\frac{1}{9}$ of 2? 4? 7? 6? 3? 12? 11?
10. What is $\frac{1}{10}$ of 7? 2? 4? 6? 9? 14? 15?
11. If 2 apples cost 3 cents what will 1 apple cost?

SOLUTION.—If 2 apples cost 3 cents, 1 apple will cost $\frac{1}{2}$ of 3 cents, which is $\frac{3}{2}$, or $1\frac{1}{2}$ cents.

12. If 2 apples cost 5 cents, what will 1 apple cost?
13. If 3 pens cost 8 cents, what will 1 pen cost?
14. If 3 yards of tape cost 14 cents, what will 1 yard cost?
15. If 5 barrels of flour cost $21, what will 1 barrel cost?
16. If 7 pecks of dried apples cost 23 dimes, what will 1 peck cost?
17. If 4 chickens cost 9 dimes, what will 1 chicken cost?
18. What will 1 pound of tamarinds cost, if 6 pounds cost 27 dimes?
19. What will 1 barrel of flour cost, if 3 barrels cost $25?
20. If you divide 7 bushels of wheat, equally, among 5 persons, how much will each receive?
21. Joshua had 13 marbles, and Lewis had $\frac{1}{2}$ as many + $\frac{1}{2}$ of a marble; how many had he?

22. A merchant divided 3 barrels of flour, equally, among 11 families; what part of a barrel did each receive?

23. A farmer divided 5 bushels of rye, equally, among 7 of his poor neighbors; what part of a bushel did he give to each?

24. Calvin had 4 pints of nuts, and shared them, equally, with 6 of his companions; how much did each receive?

25. Margaret, having 7 quarts of raspberries, shared them, equally, with 8 of her playmates; what part of a quart did each receive?

26. What will 1 pound of prunes cost, if 5 pounds cost 48 dimes?

27. What will 2 boxes of figs cost, if 7 boxes cost 29 dimes?

28. What will 4 quarts of strawberries cost, if 9 quarts cost 7 dimes?

29. What is $\frac{1}{3}$ of 2?

30. If $\frac{1}{3}$ of 2 is $\frac{2}{3}$, what is $\frac{2}{3}$ of 2?

31. What is $\frac{2}{3}$ of 4? of 5? 7? 8?

32. What is $\frac{3}{4}$ of 3? of 5? 6? 7?

33. What is $\frac{3}{5}$ of 2? of 3? 4? 9?

34. What is $\frac{5}{6}$ of 2? of 3? 5? 7?

35. What is $\frac{4}{7}$ of 3? of 5? 8? 9?

36. What is $\frac{7}{9}$ of 4? of 6? 8? 12?

37. What is $\frac{3}{8}$ of 3? of 5? 6? 9?

38. What is $\frac{3}{11}$ of 2? of 4? 5? 6?

39. What is $\frac{5}{12}$ of 2? of 4? 3? 9?

40. What is $\frac{7}{10}$ of 7? of 8? 9? 12?

41. What will $\frac{2}{3}$ of a pound of candies cost, if 1 pound costs 2 dimes?

42. What will be the cost of $\frac{3}{4}$ of a box of raisins, if 1 box cost $3?

43. What will be the cost of $\frac{3}{5}$ of a yard of cloth, at 7 dimes a yard?

44. If a ton of hay cost \$10, what will $2\frac{2}{3}$ tons cost?

45. Jeremiah is 91 years old, and $\frac{3}{7}$ of his age equals the age of his oldest son; how old is he?

46. Bought 24 yards of cloth for \$48; but, being damaged I sold $\frac{3}{4}$ of it, at $\$1\frac{5}{6}$ a yard, and the remainder for what it cost. How much did I lose?

47. Bought 14 yards of cloth; and sold $\frac{3}{7}$ of it, at \$2 a yard, which amounted to \$2 less than the whole piece cost. What did it cost a yard?

48. A horse was sold for \$97, which was $1\frac{2}{7}$ times as much as it cost. What did the horse cost?

49. If 9 yards of cloth cost \$17, what will 3 yards cost?

50. If 7 yards of cloth cost \$25, what will 9 yards cost?

51. What will 2 pounds of opium cost, if 5 pounds cost \$42?

52. If 5 pounds of indigo cost \$32, what will 2 pounds cost?

53. A wagon was sold for \$90, which was $\frac{5}{4}$ of what it cost. How much did it cost?

54. Two men started from the same place, and traveled the same way; one at the rate of 92 miles in 10 hours; the other at the rate of $1\frac{1}{10}$ miles in $\frac{1}{7}$ of an hour; how far apart will they be in 2 hours?

55. By a pipe $4\frac{5}{7}$ gallons of water run into a cistern in 1 minute; how much did the vessel hold providing it was filled in 9 minutes?

56. If 4 men can perform a certain piece of work in $13\frac{5}{7}$ days; how long would it take 7 men to do the same?

57. If 5 persons consume a barrel of flour in 9 weeks, what part of a barrel would they consume in 5 weeks?

58. If a man earn $\$\frac{7}{8}$ in a day, and a boy $\$\frac{3}{8}$, how much will they both earn in 6 days?

59. Anthony spent $\frac{3}{7}$ of all his money, and the remainder he gave for 8 yards of cloth, at $\$2\frac{3}{4}$ a yard; how much had he at first?

60. From a piece of cloth a tailor cut 5 garments, each containing $3\frac{3}{7}$ yards; and there remained $2\frac{6}{7}$ yards; how many yards did the piece contain?

61. What will 9 pounds of rice cost, if 7 pounds cost 43 cents?

62. An individual, after spending $\frac{13}{17}$ of all his money, had only $40 remaining; how much had he at first?

63. An old lady bought 30 eggs, at the rate of 2 for 5 cents; what did they cost?

64. What will 13 pounds of coffee cost, if 26 pounds cost $7?

65. What will 7 gallons of molasses cost, if 6 pints cost 27 cents?

66. If 5 lamps cost $\$7\frac{1}{2}$, what will 7 lamps cost?

67. If 5 horses can, in $4\frac{2}{5}$ days, consume 20 bushels of oats, in how many days can 11 horses consume the same?

68. If 15 gold pens cost $20, what will 5 of them cost?

69. If $\frac{7}{8}$ of an acre of land be worth $14, what are 10 acres worth?

70. $25 is $\frac{4}{7}$ of the cost of B's watch; what was the cost of his watch?

71. Mortimer's hat cost $5, and $\frac{2}{3}$ of the cost of his hat is $\frac{1}{13}$ of the cost of his coat; what was the cost of his coat?

72. If a man in $\frac{4}{15}$ of a day walk 8 miles, how far can he walk in 5 days?

73. From a piece of cloth containing 20 E. Fr., a tailor cut 8 suits of clothes, each containing $2\frac{3}{4}$ yards; how many yards remained?

74. If a man can cut 1 cord of wood in 5 hours, how many cords can he cut in 4 days, by working 12 hours a day?

75. A man bought 7 sheep, at the rate of 9 for $\$5\frac{1}{2}$; what did they cost him?

76. A boy bought 13 oranges,—giving 9 apples for 3 oranges; how many apples did his oranges cost him?

77. If 25 cents buy 7 lemons, how many cents will 9 lemons cost?

78. $\frac{3}{5}$ of 45 equals $\frac{3}{4}$ as many dollars as Andrew has; how many dollars has he?

79. $\$30\frac{5}{7}$ is $\frac{1}{7}$ of all the money A had; how much had he?

80. What will 3 pecks of flax-seed cost, if 3 pints cost 3 dimes?

81. What will 1 quart of cloverseed cost, if 2 pecks cost $3 and 2 dimes?

82. $4\frac{1}{2}$ times 7 is $\frac{1}{2}$ of what number?

83. $\frac{2}{5}$ of 36 is $\frac{4}{5}$ of what number?

84. $\frac{1}{3}$ of 36 is $\frac{12}{3}$ times what number?

85. $\frac{2}{3}$ of A's age is 3 times B's age; and B is 9 years old. What is A's age?

86. An individual, being asked the number of hours he labored each day, answered, $1\frac{1}{8}$ times the number of hours in a day is 3 times as many hours as I labor. How many hours did he labor each day?

87. $\frac{3}{5}$ of 15 is $\frac{3}{2}$ of what number?

88. $\frac{4}{7}$ of 21 is $1\frac{1}{5}$ times what number?

89. $\frac{2}{3}$ of 24 is $1\frac{3}{5}$ times what number?

90. Wright is 16 years old, and $1\frac{3}{4}$ times his age is $1\frac{2}{5}$ times Charles' age. How old is Charles?

LESSON XII.

LESSONS EIGHTH, NINTH AND TENTH COMBINED.

☞ REMARK.—*Pupils must exercise their own judgment in employing the shortest of the methods given in lessons eighth and enth, for multiplying and dividing.*

1. If 3 barrels of flour cost $\$13\frac{3}{4}$, what will 6 barrels cost?

2. If 5 pounds of opium cost $\$27\frac{1}{2}$, what will 20 pounds cost?

3. If 3 pounds of sugar cost $17\frac{3}{8}$ cents, what will 8 pounds cost?

4. If 9 pencils cost $20\frac{1}{4}$ cents, what will 12 cost?

5. How many chestnuts will pay for 9 walnuts, if 7 chestnuts are worth $10\frac{3}{7}$ walnuts?

6. If 8 barrels of flour cost $\$33\frac{3}{5}$, what will 20 barrels cost?

7. If it require $9\frac{3}{8}$ yards of cloth to make 3 coats, how many yards will it require to make 8 coats?

8. If 8 men can perform a certain piece of work in $9\frac{3}{5}$ days, how long will it take 5 men to perform the same?

9. What will be the cost of 6 sheep, if 15 cost $\$10\frac{1}{2}$?

10. If 1 person consume $10\frac{1}{26}$ bushels of wheat in a month, how much will 13 persons consume in the same time?

11. If $9\frac{7}{9}$ cents will buy 4 peaches, what will be the cost of 9 peaches?

12. If $\$9\frac{1}{16}$ will pay for 5 weeks' board, how many dollars will pay for 8 weeks' board?

13. If 6 orifices will fill a vessel in $3\frac{3}{5}$ hours, how many of the same size will be required to fill it in $\frac{1}{15}$ of an hour?

14. If 9 men can build a boat in $5\frac{3}{4}$ days, in how many days could 6 men build it?

15. If 2 men in 4 days can earn $12, how many dollars can 7 men earn in the same time?

16. If I pay $17\frac{2}{3}$ cents for riding 4 miles, how much must I pay for riding 6 miles?

17. What will 1 year's board come to, at $\$5\frac{2}{3}$ for 4 weeks?

18. If 9 barrels of fish cost $\$54\frac{1}{3}$, what will 27 barrels cost?

19. How many dollars will 1 barrel of tobacco cost if 17 barrels cost $51\frac{1}{4}$ eagles?

20 If 13 pounds of tea cost $10\frac{2}{5}$ dimes, what will 5 pounds cost?

21. If $7\frac{2}{3}$ tons of hay keep 6 horses through the winter, how many tons will keep 9 horses the same time?

22. If a fox is 40 rods before a hound and runs only 3 rods to the hound 5, how many rods will the hound run before he overtakes the fox?

23. How many dollars will a man earn in 14 days, if he earn $\$3\frac{4}{7}$ in 4 days?

24. A merchant bought 8 pieces of cloth, each piece containing 5 yards for $\$32\frac{1}{2}$; how much did it cost apiece, and how much a yard?

25. If in a certain time 6 horses eat $14\frac{3}{4}$ bushels of oats, how many bushels will 8 horses eat in the same time?

26. A boy sold 3 lemons, at the rate of 6 for 8 cents; what did he receive for them?

27. A boy gave $4\frac{1}{2}$ cents for some oranges, at the rate of 5 oranges for $7\frac{1}{2}$ cents; how many did he buy?

28. If a piece of mahogany weighing 9 pounds, is worth $\$2\frac{3}{4}$, what is the value of 12 pounds, at the same rate?

29. If a pole 8 feet long cast a shadow $4\frac{4}{5}$ feet, what will be the length of the shadow of a pole which is 15 feet long, at the same time of day?

30. At a certain time of day, a pole 5 feet long casts a shadow $7\frac{1}{2}$ feet; what is the length of that pole which, at the same time, casts a shadow $1\frac{1}{2}$ feet?

31. If it require $\$21\frac{3}{5}$ worth of provision to serve 8 men 2 days, how many dollar's worth will serve 5 men 1 day?

32. What is the length of a pole the shadow of which is 12 feet long, at the same time, a pole $2\frac{2}{3}$ feet in length, casts a shadow 4 feet long?

LESSON XIII.

Remark.—By inspecting *Lessons* 8th and 10th, we observe, that, multiplying both *numerator* and *denominator* by the same number, does not alter the value of the fraction. Hence, to convert a fraction to an equivalent fraction having a different denominator, we may multiply both numerator and denominator by any number which will cause the fraction to have the required denominator.

1. James gave his sister $\frac{1}{2}$ of an apple; how many fourths was that?

2. $\frac{1}{3}$ is how many sixths?

3. Byron gave his sister $\frac{1}{2}$ of an apple, and his brother $\frac{1}{4}$ of an apple; how many fourths did he give away?

4. $\frac{1}{2}$ is how many sixths?
5. $\frac{1}{3}$ is how many sixths?
6. $\frac{1}{2}$ and $\frac{1}{3}$ are how many sixths?
7. $\frac{2}{3}$ are how many sixths?
8. $\frac{5}{6}$ are how many twelfths?
9. $\frac{2}{6}$ are how many twelfths?
10. $\frac{3}{6}$ are how many eighteenths?
11. $\frac{3}{5}$ are how many tenths?
12. $\frac{2}{5}$ are how many tenths?
13. $\frac{4}{5}$ are how many twentieths?
14. $\frac{1}{2}$ is how many tenths?

15. Hafford gave $\frac{3}{5}$ of an orange to his sister; how many fifteenths did he give away?

16. How many sixteenths in $\frac{3}{8}$?
17. How many sixteenths in $\frac{5}{8}$?
18. How many sixteenths in $\frac{7}{8}$?
19. How many fourteenths in $\frac{3}{7}$?
20. How many fourteenths in $\frac{5}{7}$?
21. How many fourteenths in $\frac{4}{7}$?
22. How many ninths in $\frac{2}{3}$?
23. How many twentieths in $\frac{4}{5}$?
24. How many fortieths in $\frac{7}{8}$?
25. How many forty-ninths in $\frac{4}{7}$?

26. How many fifteenths in $\frac{3}{5}$

27. A man gave $\frac{1}{5}$ of a bushel of potatoes to one poor woman, and $\frac{3}{10}$ of a bushel to another; what part of a bushel did he give to both?

28. How could you divide an apple so as to give $\frac{3}{8}$ of it to one boy, and $\frac{1}{4}$ of it to another?

29. $\frac{1}{2} + \frac{2}{3}$ are how many sixths?

30. $\frac{3}{4} + \frac{2}{3}$ are how many twelfths?

31. $\frac{8}{9} + \frac{1}{2}$ are how many eighteenths?

32. $\frac{7}{8} + \frac{2}{3}$ are how many twenty-fourths?

33. $\frac{3}{7}$ are how many times $\frac{3}{21}$?

34. What is the sum of $\frac{3}{4}$ and $\frac{3}{8}$?

35. What is the sum of $\frac{3}{7}$ and $\frac{2}{5}$?

36. What is the sum of $\frac{1}{3}$ and $\frac{3}{7}$?

37. What is the sum of $\frac{2}{3}$ and $\frac{3}{5}$?

38. What is the sum of $\frac{3}{7}$ and $\frac{7}{8}$?

39. What is the sum of $\frac{1}{2}$ and $\frac{2}{3}$?

40. What is the sum of $\frac{5}{9}$ and $\frac{2}{3}$?

41. What is the sum of $\frac{7}{9}$ and $\frac{1}{2}$?

42. What is the sum of $\frac{4}{5}$ and $\frac{4}{7}$?

43. What is the sum of $\frac{5}{6}$, $\frac{1}{2}$ and $\frac{2}{3}$?

44. What is the sum of $\frac{3}{4}$, $\frac{2}{3}$ and $\frac{5}{6}$?

45. From $\frac{3}{4}$ subtract $\frac{1}{2}$.

46. From $\frac{5}{6}$ subtract $\frac{2}{3}$.

47. From $\frac{5}{7}$ subtract $\frac{2}{3}$.

48. From $2\frac{1}{3}$ subtract $\frac{3}{7}$.

49. From 4 subtract $\frac{1}{2}$.

50. From 3 subtract $\frac{3}{5}$.

51. From 9 subtract $\frac{3}{7}$.

52. From 5 subtract $\frac{4}{5}$.

53. From 3 subtract $1\frac{1}{2}$.

54. From 9 subtract $2\frac{2}{3}$.

55. From 6 subtract $3\frac{3}{4}$.

56. $14 - 3\frac{1}{3}$ are how many?
57. $7 - 2\frac{5}{6}$ are how many?
58. $9 - 3\frac{4}{7}$ are how many?
59. $10 - 3\frac{5}{8}$ are how many?
60. $12 - 3\frac{4}{9}$ are how many?
61. $13 - 7\frac{2}{11}$ are how many?
62. $9\frac{2}{3} - 4\frac{1}{2}$ are how many?
63. $7\frac{2}{3} - 5\frac{3}{7}$ are how many?
64. $3\frac{1}{2} - 1\frac{4}{5}$ are how many?
65. $4\frac{3}{7} - 1\frac{5}{6}$ are how many?
66. $5\frac{3}{5} - 2\frac{2}{3}$ are how many?
67. $9\frac{3}{5} - 7\frac{2}{3}$ are how many?
68. $2\frac{3}{4} + 3\frac{1}{4} - \frac{3}{5}$ are how many?
69. $4\frac{2}{3} + 5\frac{3}{4} - 2\frac{3}{5}$ are how many?
70. $3\frac{5}{6} + 4\frac{1}{6} - \frac{7}{8}$ are how many?
71. $9\frac{2}{3} + 3\frac{3}{4} - 3$ are how many?
72. $\frac{2}{3} + \frac{3}{4} - \frac{7}{8}$ are how many?
73. $\frac{5}{6} + \frac{1}{3} + \frac{5}{7} - \frac{1}{2}$ are how many?
74. $2\frac{1}{8}$ are how many times $\frac{2}{18}$?
75. $3\frac{1}{5}$ are how many times $\frac{4}{15}$?
76. $\frac{9}{13}$ are how many times $\frac{3}{26}$?
77. $\frac{1}{2}$ is how many times $\frac{6}{36}$?
78. $\frac{4}{7}$ are how many times $\frac{2}{14}$?
79. $1\frac{1}{8}$ are how many times $\frac{3}{16}$?
80. $\frac{7}{9}$ are how many times $\frac{2}{18}$?
81. $\frac{5}{13}$ are how many times $\frac{5}{26}$?
82. $8\frac{2}{3}$ are how many times $2\frac{1}{6}$?
83. $10\frac{4}{5}$ are how many times $\frac{9}{15}$?
84. $12\frac{1}{2}$ are how many times $1\frac{1}{4}$?
85. $\frac{1}{6} + \frac{1}{5} + \frac{1}{4}$ lacks how much of being a whole one?
86. $\frac{1}{7} + \frac{2}{5} + \frac{2}{7}$ lacks how much of being a whole one?

87. A lady gave $\frac{1}{2}$ of all her money for a dress, and $\frac{3}{7}$ of it for a shawl; what part of her money had she remaining?

88. $\frac{1}{2}$ of an army was killed, and $\frac{1}{3}$ taken prisoners; what part of the army escaped?

89. $\frac{3}{8}$ of an army was killed, $\frac{5}{9}$ taken prisoners, and 500 escaped unhurt; how many were there in the army?

90. $\frac{2}{3}$ of the length of a pole is in the ground, $\frac{1}{5}$ in the water, and 12 feet out of the water; what is the length of the pole?

91. A market woman sold $\frac{5}{9}$ of all her oranges to one man and $\frac{1}{3}$ of them to another, and then had only 9 remaining; how many had she at first, and how many did she sell to each?

92. A man, after spending $\frac{3}{4}$ of his fortune, found that \$20 was $\frac{2}{9}$ of what he had remaining; what was his fortune?

93. A hawk caught $\frac{2}{5}$ of Euphemia's chickens, a cat killed $\frac{1}{3}$ of them, and $\frac{1}{7}$ of them died; she then had 13 remaining; how many had she at first, and how many were destroyed by the hawk and cat respectively?

94. Said A to B if to my age you add its $\frac{1}{2}$ and $\frac{2}{5}$, the sum will be 38; how old was he?

95. A is 40 years old, and $\frac{3}{4}$ of his age is $\frac{3}{5}$ of twice as much as his wife's age; how old was his wife?

LESSON XIV.

REMARK.—By inspecting *Lessons* 8th and 10th, we observe, that dividing both *Numerator* and *Denominator* by the same number, does not alter the value of the fraction. Hence to reduce a fraction to its lowest terms, we may divide both numerator and denominator by any number that is contained in each of them without a remainder.

1. Reduce $\frac{4}{8}$ to its lowest terms.

2. Reduce $\frac{3}{9}$ to its lowest terms.

3. Reduce $\frac{3}{6}$ to its lowest terms.
4. Reduce $\frac{5}{15}$ to its lowest terms.
5. Reduce $\frac{4}{12}$ to its lowest terms.
6. Reduce $\frac{12}{24}$ to its lowest terms.
7. Reduce $\frac{15}{25}$ to its lowest terms.
8. Reduce $\frac{45}{75}$ to its lowest terms.
9. Reduce $\frac{36}{48}$ to its lowest terms.
10. Reduce $\frac{25}{35}$ to its lowest terms.
11. Reduce $\frac{36}{60}$ to its lowest terms.
12. Reduce $\frac{75}{100}$ to its lowest terms
13. Reduce $\frac{32}{56}$ to its lowest terms.
14. Reduce $\frac{50}{70}$ to its lowest terms.
15. Reduce $\frac{12}{18}$ to its lowest terms.
16. Why does the value of the fraction remain the same, when you divide both *numerator* and *denominator* by the same number ?
17. When you multiply both numerator and denominator by the same number, why does it not change the value of the fraction ?
18. Reduce 4 times $\frac{3}{4}$ to its lowest terms.
19. Reduce 7 times $\frac{4}{14}$ to its lowest terms.
20. Reduce 8 times $2\frac{1}{4}$ to its lowest terms.
21. Reduce 6 times $\frac{5}{24}$ to its lowest terms.
22. Reduce 4 times $\frac{5}{15}$ to its lowest terms.
23. Reduce 12 times $\frac{3}{18}$ to its lowest terms.
24. Reduce 12 times $\frac{5}{48}$ to its lowest terms.
25. Reduce 8 times $\frac{2}{24}$ to its lowest terms.
26. Reduce 7 times $\frac{3}{05}$ to its lowest terms.
27. Reduce 5 times $\frac{14}{35}$ to its lowest terms.
28. Reduce 4 times $\frac{3}{36}$ to its lowest terms.
29. Reduce 6 times $\frac{18}{36}$ to its lowest terms.
30. Reduce 9 times $\frac{27}{81}$ to its lowest terms.

LESSON XV.

1. If you cut an apple into two equal pieces, what will 1 of these pieces be called?

2. If you cut $\frac{1}{2}$ of an apple into two equal pieces, what part of a whole apple will 1 of these pieces be called?

3. If Alice has $\frac{1}{3}$ of a lemon, and gives $\frac{1}{2}$ of it to Ann, what part of a lemon will Ann receive?

SOLUTION.—$\frac{1}{3}$ equals $\frac{2}{6}$. Therefore, Ann receives $\frac{1}{2}$ of $\frac{2}{6}$ of a lemon, which is $\frac{1}{6}$ of a lemon.

4. George, having $\frac{1}{2}$ of a melon, gave $\frac{1}{2}$ of it to Marcus; what part of a melon did he receive?

5. Crary had $\frac{1}{4}$ of a dollar, and gave $\frac{1}{2}$ of it to Joshua; what part of a dollar did Joshua receive?

6. Robert had $\frac{1}{5}$ of a dollar, and gave $\frac{1}{2}$ of it for a cake; what did the cake cost him?

7. Margaret had $\frac{1}{6}$ of a pound of candies, and Mary had $\frac{1}{2}$ as much; how much had Mary?

8. Jane had $\frac{1}{5}$ of a pound of sugar, and Ann $\frac{1}{3}$ as much; how much had she?

9. A boy bought $\frac{1}{6}$ of a quart of chestnuts, and gave $\frac{1}{3}$ of them to his sister; what part of a quart did she receive?

10. A man owned $\frac{1}{9}$ of a share in a bank, and sold $\frac{1}{2}$ of that; what part of a share had he remaining?

11. B owned $\frac{1}{8}$ of a ship, and sold $\frac{1}{4}$ of his share · what part of a whole ship did he sell?

12. What is $\frac{1}{3}$ of $\frac{1}{5}$?
13. What is $\frac{1}{5}$ of $\frac{1}{2}$?
14. What is $\frac{1}{3}$ of $\frac{1}{7}$?
15 What is $\frac{1}{4}$ of $\frac{1}{6}$?
16 What is $\frac{1}{8}$ of $\frac{1}{9}$?
17 What is $\frac{1}{7}$ of $\frac{1}{9}$?
18. What is $\frac{1}{7}$ of $\frac{1}{4}$?
19. What is $\frac{1}{2}$ of $\frac{1}{8}$?

20. What is $\frac{1}{11}$ of $\frac{1}{3}$?

21. A kite was up in the air and fell $\frac{7}{8}$ of the way to the ground, it then arose $\frac{1}{9}$ of the distance it lacked of being to the ground; what part of the whole distance was it above the ground?

22. Homer is $\frac{1}{5}$ as old as his father, and Nelson is $\frac{1}{9}$ as old as Homer; what part of the father's age is Nelson's age?

23. A man, owning $\frac{1}{4}$ of a barrel of fish, accommodated his neighbor with $\frac{1}{7}$ of it; how much had he remaining?

24. A man, having $\frac{1}{2}$ of an eagle, gave $\frac{1}{5}$ of it to B, and B gave $\frac{1}{10}$ of what he had to C; what part of an eagle had each after this division, and how many cents had each?

25. Elizabeth had $\frac{2}{3}$ of a pie, and gave $\frac{1}{3}$ of it to Harriet; how much did she receive?

SOLUTION.—$\frac{1}{3}$ of $\frac{1}{3}$ is $\frac{1}{9}$; and if $\frac{1}{3}$ of $\frac{1}{3}$ is $\frac{1}{9}$, $\frac{1}{3}$ of $\frac{2}{3}$ is twice $\frac{1}{9}$, which are $\frac{2}{9}$. Therefore Harriet had $\frac{2}{9}$ of a pie.

26. What is $\frac{1}{2}$ of $\frac{3}{5}$?

27. What is $\frac{1}{4}$ of $\frac{2}{3}$?

28. What is $\frac{1}{5}$ of $\frac{2}{3}$?

29. What is $\frac{1}{5}$ of $\frac{3}{4}$?

30. What is $\frac{1}{7}$ of $\frac{3}{4}$?

31. What is $\frac{1}{6}$ of $\frac{3}{5}$?

32. What is $\frac{1}{9}$ of $\frac{2}{3}$?

33. What is $\frac{1}{9}$ of $\frac{4}{5}$?

34. What is $\frac{1}{8}$ of $\frac{3}{6}$?

35. What is $\frac{1}{5}$ of $\frac{3}{5}$?

36. What is $\frac{1}{4}$ of $\frac{9}{13}$?

37. What is $\frac{1}{9}$ of $\frac{3}{8}$?

38. What is $\frac{1}{7}$ of $\frac{6}{7}$?

39. What is $\frac{1}{7}$ of $1\frac{3}{7}$

40. What is $\frac{3}{5}$ of $\frac{1}{4}$?

41. What is $\frac{3}{4}$ of $\frac{1}{5}$?
42. What is $\frac{3}{4}$ of $\frac{1}{7}$?
43. What is $\frac{3}{5}$ of $\frac{1}{9}$?
44. What is $\frac{5}{6}$ of $\frac{1}{7}$?
45. What is $\frac{3}{7}$ of $\frac{1}{5}$?
46. What is $\frac{3}{8}$ of $\frac{1}{9}$?
47. What is $\frac{7}{8}$ of $\frac{1}{4}$?
48. What is $\frac{5}{7}$ of $\frac{1}{8}$?
49. What is $\frac{4}{5}$ of $\frac{1}{6}$?
50. What is $\frac{3}{7}$ of $\frac{1}{4}$?
51. What is $\frac{2}{3}$ of $\frac{3}{5}$?
52. What is $\frac{3}{4}$ of $\frac{2}{5}$?
53. What is $\frac{3}{5}$ of $\frac{2}{7}$?
54. What is $\frac{3}{4}$ of $\frac{2}{8}$?
55. What is $\frac{3}{4}$ of $\frac{3}{5}$?
56. What is $\frac{3}{8}$ of $\frac{5}{7}$?
57. What is $\frac{3}{5}$ of $\frac{6}{7}$?
58. What is $\frac{4}{7}$ of $\frac{8}{9}$?
59. What is $\frac{5}{7}$ of $\frac{6}{7}$?
60. What is $\frac{8}{13}$ of $\frac{2}{3}$?
61. What part of 1 is $\frac{2}{3}$ of $\frac{1}{5}$?
62. What part of 1 is $\frac{3}{7}$ of $\frac{1}{6}$?
63. What part of 2 is $\frac{4}{6}$ of $\frac{1}{2}$?
64. What part of 2 is $\frac{1}{4}$ of $1\frac{4}{5}$?
65. What part of 2 is $\frac{3}{4}$ of $\frac{1}{3}$?
66. What part of 3 is $\frac{1}{2}$ of $\frac{1}{2}$?
67. What part of 4 is $\frac{1}{3}$ of $1\frac{4}{5}$?
68. What part of 5 is $\frac{1}{5}$ of $\frac{2}{3}$?
69. What part of 9 is $\frac{3}{7}$ of $1\frac{1}{3}$?
70. What part of 2 is $\frac{2}{8}$ of $\frac{3}{4}$?
71. What part of 2 is $\frac{3}{1}$ of $2\frac{1}{3}$?
72. What part of 4 is $\frac{3}{5}$ of $\frac{3}{15}$?

73. What part of 6 is $\frac{2}{7}$ of $\frac{5}{6}$?

74. What part of 3 is $\frac{1}{2}$ of $4\frac{2}{5}$?

75. What part of 4 is $\frac{5}{6}$ of $12\frac{5}{6}$?

76. What part of 7 is $\frac{2}{3}$ of $10\frac{3}{4}$?

77. Anthony had $\frac{1}{8}$ of $\frac{3}{4}$ of a pound of cinnamon; what part of a pound had he?

78. Albert had $\frac{1}{7}$ of $\frac{2}{3}$ of a quart of strawberries; how many strawberries had he, providing 1 quart contained 42 strawberries?

79. Abner gave $\frac{1}{7}$ of $\frac{6}{7}$ of a melon to his brother; what part of a melon had he remaining?

80. Matilda bought $\frac{5}{6}$ of a quart of milk for tea, and spilled $\frac{1}{3}$ of it; what part of a quart had she remaining?

81. Edwin picked $\frac{7}{8}$ of a pail-full of blackberries, and on his way home spilled $\frac{1}{3}$ of them; what part of a pail-full had he remaining?

82. A merchant bought $\frac{4}{5}$ of a hogshead of molasses, and $\frac{1}{5}$ of it leaked out; what part of a hogshead had he remaining?

83. Hannah had $\frac{2}{3}$ of a pound of candies, and gave $\frac{3}{4}$ of them to Augusta; what part of a pound did she give Augusta?

84. Elisha found $\$\frac{3}{4}$ and gave $\frac{2}{5}$ of it to Ephraim; what part of a dollar had he remaining?

85. Andrew bought $\frac{4}{5}$ of a pound of maple-sugar, nd gave $\frac{2}{11}$ of it to Walter; what part of a pound did Valter receive?

86. Jacob, having a pine-apple, gave $\frac{2}{3}$ of $\frac{5}{7}$ of it the one that could tell how much that would be; ow much was that individual to receive?

87. James gave $\frac{2}{3}$ of $\frac{3}{5}$ of a dime for a top; what ld the top cost him?

88. Robert gave $\frac{3}{4}$ of a dollar for a cap; what did he cap cost him?

89. Mary gave $\frac{7}{8}$ of $1\frac{3}{5}$ dimes for a comb; what did the comb cost her?

90. Clorinda gave $\frac{2}{3}$ of 6 dimes for a pair of gloves; how much did the gloves cost her?

91. A man, having $4\frac{3}{5}$ barrels of flour, sold $\frac{2}{3}$ of it; how much remained unsold?

92. A man gave $\frac{3}{5}$ of $\$3\frac{1}{3}$ for a silver pencil; what was the cost of the pencil?

93. Jane worked $8\frac{3}{4}$ hours in a day, and Delilah worked only $\frac{4}{5}$ as many; how many hours did she work in a day?

94. B gave $\$32\frac{2}{3}$ for a cow, which was $\frac{2}{3}$ as much as A gave for his; what did A's cow cost him?

95 Darius is $18\frac{3}{4}$ years old, and Daniel is $\frac{3}{4}$ as old; how old is he?

96. If 1 yard of cloth cost $\$5\frac{3}{4}$, what will $\frac{2}{3}$ of a yard cost?

97. If 4 yards of cloth cost $\$9\frac{1}{2}$, what will $\frac{4}{5}$ of a yard cost?

98. If 5 barrels of beef cost $\$18\frac{1}{2}$, what will $\frac{1}{2}$ of a barrel cost?

99. If $\frac{2}{3}$ of an apple cost $\frac{3}{4}$ of a cent, what will 1 apple cost.

100. If $\frac{1}{4}$ of an orange cost $1\frac{1}{3}$ cents, what will $\frac{5}{16}$ of an orange cost?

101. If 5 pounds of butter cost $6\frac{1}{4}$ dimes, how many cents will $1\frac{3}{5}$ pounds cost?

LESSON XVI.

1. If 4 barrels of flour cost $\$14\frac{2}{5}$, what will $\frac{5}{6}$ of a barrel cost?

2. If 3 bushels of potatoes cost $5\frac{1}{7}$ dimes, what will $1\frac{1}{6}$ bushels cost?

3. If $2\frac{1}{2}$ bushels of apples cost $6\frac{1}{4}$ dimes, how many cents will $\frac{4}{5}$ of a bushel cost?

4. If $\frac{3}{4}$ of an apple cost $\frac{2}{3}$ of a cent, what will 1 apple cost?

5. If $\frac{3}{5}$ of an orange cost $\frac{3}{4}$ of a cent, what will $\frac{3}{4}$ of an orange cost?

6. If $2\frac{3}{4}$ yards of cassimere cost $\$3\frac{5}{6}$, what will $5\frac{1}{2}$ yards cost?

7. If $5\frac{2}{5}$ yards of shalloon cost $\$5\frac{4}{10}$, what will 2 yards cost?

8. If in $3\frac{4}{7}$ hours A can do a certain piece of work, how long will it take him to do a piece of work $1\frac{2}{5}$ times as large?

9. $\frac{3}{4}$ of A's age is $\frac{2}{3}$ of B's; and $\frac{3}{4}$ of B's age is $\frac{2}{3}$ of C's age. How old are A and B respectively, providing C is 81 years old?

10. Bought $3\frac{2}{3}$ boxes of goods, at $\$6\frac{6}{11}$ a box, and paid for it with sheep, at \$2 a head; how many sheep did it take?

11. How many times $\frac{2}{5}$ is $\frac{3}{4}$?

Solution.—1 is contained in $\frac{3}{4}$, $\frac{3}{4}$ times; and, if 1 is contained in $\frac{3}{4}$, $\frac{3}{4}$ times, $\frac{1}{5}$ is contained in $\frac{3}{4}$ 5 times $\frac{3}{4}$ times, which are $\frac{15}{4}$ times, and $\frac{2}{5}$ is contained, $\frac{1}{2}$ of $\frac{15}{4}$ times, which is $\frac{15}{8}$ times.

12. How many times $\frac{2}{3}$ is $\frac{3}{4}$?

13. How many times $\frac{2}{3}$ is $\frac{7}{8}$?

14. How many times $\frac{3}{5}$ is $\frac{6}{7}$?

15. How many times $\frac{3}{8}$ is $\frac{3}{16}$?

16. How many times $\frac{2}{3}$ is $1\frac{1}{3}$?

17. How many times $\frac{2}{7}$ is $\frac{4}{14}$?

18. How many times $\frac{2}{5}$ is $\frac{7}{8}$?

19. How many times $\frac{4}{5}$ is $3\frac{3}{7}$?

20. How many times $\frac{3}{4}$ is $2\frac{2}{3}$?

21. How many times $\frac{3}{7}$ is $2\frac{2}{5}$?

22. How many times $\frac{5}{9}$ is $5\frac{5}{9}$?

23. How many times $\frac{2}{5}$ is $\frac{3}{10}$?

24. How many times $\frac{4}{7}$ is $\frac{12}{14}$?

25. How many times $\frac{5}{6}$ is $3\frac{1}{3}$?

26. How many times $\frac{7}{8}$ is $4\frac{1}{2}$?

27. How many times $1\frac{1}{2}$ is $2\frac{1}{3}$?

28. How many times $2\frac{2}{3}$ is $1\frac{4}{5}$?

29. How many times $3\frac{1}{4}$ is $5\frac{1}{5}$?

30. How many times $4\frac{2}{3}$ is $5\frac{3}{5}$?

31. A farmer sold a quantity of rye for $96, which was only $\frac{4}{5}$ of what it was worth; how much did he lose by the bargain ?

32. A man sold a cow for $1\frac{3}{8}$ times what she cost him, and by so doing gained $6; what did the cow cost him ?

33. A merchant sold a quantity of goods for $\frac{13}{18}$ of what they cost, and by so doing, he lost $15; what did the goods cost him ?

34. A farmer, having lost 12 sheep, found, that only $\frac{7}{8}$ of his flock remained; how many sheep had he remaining ?

35. An individual, being asked how many geese he had, answered, that if to $\frac{5}{7}$ of his flock 24 geese were added, the sum would equal $1\frac{2}{7}$ times his original flock; how many geese had he ?

36. If $\frac{2}{3}$ of a yard of cloth cost $\$\frac{3}{4}$, what will $\frac{2}{5}$ of a yard cost ?

37. A boy, being asked his age, said, that $8\frac{1}{4}$ years was $\frac{3}{8}$ of twice as much as his age; how old was he ?

38. If $\frac{3}{5}$ of the candies I have cost $7\frac{1}{2}$ cents, what will $\frac{7}{8}$ of them cost ?

39. What will $\frac{5}{6}$ of a barrel of flour cost, if $\frac{5}{7}$ of a barrel cost $\$2\frac{1}{7}$?

40. What will $\frac{2}{3}$ of an orange cost, if $\frac{5}{7}$ of an orange cost $2\frac{1}{2}$ cents ?

41. How many yards of cloth will be required to make a coat, if $1\frac{2}{3}$ yards will make $\frac{2}{3}$ of a coat ?

42. $\frac{3}{4}$ of 2 are how many times $\frac{2}{3}$?

43. $\frac{3}{16}$ of 8 are how many times $\frac{1}{2}$?

44. $\frac{3}{5}$ of 7 are how many times 3 ?

45. $\frac{3}{5}$ of 8 are how many times $\frac{2}{5}$?

46. $\frac{3}{4}$ of 12 are how many times $\frac{4}{3}$ of 6

47. $\frac{2}{5}$ of 7 are how many times $\frac{3}{5}$ of 2 ?

48. If $\frac{2}{5}$ of 3 yards of cloth cost \$$1\frac{1}{5}$, what will $\frac{3}{5}$ of 7 yards cost ?

49. If $\frac{4}{7}$ of 6 yards of cloth cost \$$2\frac{3}{7}$, how much will $\frac{3}{5}$ of 7 yards cost ?

50. If $\frac{2}{3}$ of $\frac{3}{4}$ of a barrel of flour cost \$$1\frac{2}{3}$, what will $\frac{1}{2}$ of $\frac{2}{3}$ of $\frac{3}{4}$ of a barrel cost ?

LESSON XVII.

1. 12 is $\frac{3}{4}$ of what number ?

SOLUTION.—If 12 is $\frac{3}{4}$ of some number, $\frac{1}{4}$ of that number is $\frac{1}{3}$ of 12, which is 4 ; if 4 is $\frac{1}{4}$ of that number, $\frac{4}{4}$ (which is that number) is 4 times 4, which are **16**. **Therefore 12 is** $\frac{3}{4}$ of 16.

2. 15 is $\frac{3}{5}$ of what number ?
3. 18 is $\frac{2}{7}$ of what number ?
4. 20 is $\frac{4}{5}$ of what number ?
5. 26 is $\frac{2}{3}$ of what number ?
6. 25 is $\frac{5}{7}$ of what number ?
7. 30 is $\frac{5}{8}$ of what number ?
8. 32 is $\frac{8}{9}$ of what number ?
9. 36 is $\frac{6}{7}$ of what number ?
10. 36 is $\frac{6}{11}$ of what number ?
11. 36 is $\frac{9}{2}$ of what number ?
12. 24 is $\frac{6}{7}$ of what number ?
13. 9 is $\frac{3}{7}$ of what number ?
14. 12 is $\frac{4}{5}$ of what number ?
15. 38 is $\frac{2}{3}$ of what number ?
16. 16 is $\frac{4}{5}$ of what number ?
17. 16 is $\frac{4}{7}$ of what number ?
18. 16 is $\frac{2}{3}$ of what number ?

19. 40 is $\frac{5}{8}$ of what number ?
20. 40 is $\frac{8}{9}$ of what number ?
21. 72 is $\frac{8}{9}$ of what number ?
22. 72 is $\frac{9}{10}$ of what number ?
23. 12 is $\frac{2}{3}$ of how many times 2 ?
24. 16 is $\frac{4}{9}$ of how many times 3 ?
25. 18 is $\frac{2}{5}$ of how many times 9 ?
26. 32 is $\frac{4}{7}$ of how many times 4 ?
27. 46 is $\frac{2}{3}$ of how many times 23 ?
28. 48 is $\frac{4}{5}$ of how many times 5 ?
29. 48 is $\frac{4}{7}$ of how many times 4 ?
30. 36 is $\frac{6}{7}$ of how many times 2 ?
31. 30 is $\frac{5}{6}$ of how many times $\frac{1}{2}$ of 12 ?
32. 30 is $\frac{6}{7}$ of how many times $\frac{1}{2}$ of 10 ?
33. 16 is $\frac{4}{9}$ of how many times $\frac{2}{3}$ of 9 ?
34. 16 is $\frac{4}{15}$ of how many times $\frac{3}{4}$ of 16 ?
35. 24 is $\frac{3}{8}$ of how many times $\frac{2}{3}$ of 12 ?
36. 25 is $\frac{5}{9}$ of how many times $\frac{1}{3}$ of 9 ?
37. 35 is $\frac{5}{12}$ of how many times $\frac{2}{3}$ of 9 ?
38. 40 is $\frac{5}{8}$ of how many times $\frac{4}{5}$ of 10 ?
39. 48 is $\frac{6}{10}$ of how many times $\frac{4}{5}$ of 25 ?
40. 96 is $\frac{2}{3}$ of how many times $\frac{3}{4}$ of 16 ?

LESSON XVIII.

1. $\frac{2}{3}$ of 6 is $\frac{2}{5}$ of what number ?
2. $\frac{2}{5}$ of 10 is $\frac{2}{3}$ of what number ?
3. $\frac{3}{4}$ of 8 is $\frac{2}{7}$ of what number ?
4. $\frac{2}{7}$ of 21 is $\frac{3}{4}$ of what number ?
5. $\frac{4}{5}$ of 15 is $\frac{3}{10}$ of what number ?
6. $\frac{3}{10}$ of 40 is $\frac{4}{5}$ of what number ?

7. $\frac{2}{9}$ of 27 is $\frac{2}{5}$ of what number?
8. $\frac{5}{9}$ of 27 is $\frac{5}{13}$ of what number?
9. $\frac{4}{9}$ of 81 is $\frac{9}{10}$ of what number?
10. $\frac{6}{7}$ of 49 is $\frac{6}{11}$ of what number?
11. $\frac{2}{3}$ of 12 is $\frac{1}{2}$ of how many times 2?
12. $\frac{3}{4}$ of 16 is $\frac{2}{3}$ of how many times 2?
13. $\frac{4}{5}$ of 10 is $\frac{2}{7}$ of how many times 4?
14. $\frac{1}{4}$ of 16 is $\frac{2}{15}$ of how many times 6?
15. $\frac{4}{5}$ of 15 is $\frac{2}{7}$ of how many times 6?
16. $\frac{2}{5}$ of 20 is $\frac{2}{3}$ of how many times 3?
17. $\frac{5}{6}$ of 12 is $\frac{2}{9}$ of how many times 5?
18. $\frac{3}{4}$ of 20 is $\frac{5}{9}$ of how many times 3?
19. $\frac{7}{6}$ of 36 is $\frac{21}{10}$ of how many times 4?
20. $\frac{8}{9}$ of 72 is $\frac{2}{3}$ of how many times 12?
21. $\frac{9}{8}$ of 96 is $\frac{2}{5}$ of how many times 90?
22. $\frac{7}{9}$ of 117 is $\frac{7}{8}$ of how many times 4?
23. $\frac{5}{7}$ of 56 is $\frac{5}{6}$ of how many times 8?
24. $\frac{5}{6}$ of 60 is $\frac{2}{3}$ of how many times 5?
25. $\frac{2}{3}$ of 36 is $\frac{2}{5}$ of how many times 12?
26. $\frac{3}{4}$ of 72 is $\frac{9}{8}$ of how many times 5?
27. $\frac{7}{8}$ of 40 is $\frac{5}{12}$ of how many times 21?
28. $\frac{5}{8}$ of 32 is $\frac{4}{9}$ of how many times 9?
29. $\frac{2}{3}$ of 15 is $\frac{5}{7}$ of how many times 2?
30. $\frac{3}{5}$ of 15 is $\frac{1}{3}$ of how many times 9?
31. $\frac{5}{8}$ of 24 is $\frac{5}{4}$ of how many times 3?
32. $\frac{7}{9}$ of 45 is $\frac{5}{6}$ of how many times 3?
33. $\frac{5}{7}$ of 14 is $\frac{2}{9}$ of how many times 5?
34. $\frac{7}{9}$ of 18 is $\frac{2}{9}$ of how many times 7?
35. $\frac{9}{10}$ of 40 is $\frac{3}{8}$ of how many times 6?
36. $\frac{4}{5}$ of 45 is $\frac{6}{11}$ of how many times 3?
37. $\frac{6}{7}$ of 35 is $\frac{5}{9}$ of how many times 2?
38. $\frac{5}{9}$ of 81 is $\frac{5}{9}$ of how many times 9?

39. $\frac{2}{3}$ of 5 is $\frac{5}{7}$ of how many times 7?

40. $\frac{3}{4}$ of 7 is $\frac{3}{5}$ of how many times 3?

41. B's horse cost \$60, and $\frac{4}{5}$ of the cost of the horse, is $\frac{1}{4}$ of two times the value of his wagon; what is the value of his wagon?

42. A coat cost \$20, and $\frac{4}{5}$ of the cost of the coat, is $\frac{2}{7}$ of 8 times the price of a hat; the price of the hat is required.

43. If a cow cost \$30, and $\frac{2}{3}$ of this, is $\frac{2}{7}$ of 10 times the price of a sheep; what is the price of a sheep?

44. A's farm is worth \$1200, and $\frac{4}{5}$ of its value, is $\frac{3}{8}$ of 10 times the value of its yearly productions; what is the value of the yearly productions?

45. The articles contained in a certain store cost \$500, and $\frac{3}{10}$ of their cost, is $\frac{2}{5}$ of 3 times the amount paid for the silks; what did the silks cost, and what the other articles?

46. A's wedding clothes cost \$180, and $\frac{2}{3}$ of the cost of his clothes, is $\frac{2}{9}$ of 6 times the cost of his wife's wedding dress; what was the cost of her dress?

47. The insurance of a ship amounted to \$800, and $\frac{1}{4}$ of that is $\frac{1}{20}$ of 2 times the value of the cargo; what is the value of the cargo?

48. A's house cost \$1400, and $\frac{7}{4}$ of its cost, is $3\frac{1}{3}$ times $\frac{1}{2}$ of the cost of the furniture contained in it; what was the cost of the furniture?

49. Providing a house was worth \$1200, and $\frac{3}{4}$ of its value was $\frac{2}{3}$ of $\frac{1}{2}$ times the value of the farm on which it stood; what was the value of the farm?

50. If a sleigh cost \$100, what would be the cost of a wagon, if $\frac{2}{5}$ of the cost of the sleigh was $\frac{2}{11}$ of twice the cost of a wagon?

51. Lambert is worth \$2500, and $\frac{4}{5}$ of his fortune, is $3\frac{1}{3}$ times $\frac{1}{2}$ of Latham's fortune; how much is Latham worth?

Distances on the Railroad between Albany and Buffalo.

52. The distance from Albany to Schenectady, is

16 miles, and $\frac{3}{4}$ of this distance, is $\frac{2}{3}$ of $\frac{1}{6}$ times the distance from Albany to Rome; what is the distance to Rome?

53. Fort-Plain is 56 miles from Albany, and $\frac{5}{7}$ of this distance is $1\frac{3}{5}$ times $\frac{1}{10}$ of the distance from Albany to Rochester; what is the distance to Rochester?

54. Waterloo is 192 miles from Albany, and $\frac{5}{6}$ of this distance is $1\frac{2}{3}$ times the distance to Utica, and 3 miles more; what is the distance to Utica?

55. Buffalo is 325 miles from Albany, and $\frac{3}{5}$ of this distance is $7\frac{1}{2}$ times $\frac{1}{11}$ of the distance to Batavia, and 5 miles more; what is the distance to Batavia?

Distances on the Railroad between Albany and Boston.

56. Boston is 200 miles from Albany, and $\frac{2}{5}$ of this distance, is $1\frac{3}{5}$ times $\frac{1}{2}$ of the distance to West Springfield; what is the distance to West Springfield?

57. From Albany to the state line is 38 miles, and $1\frac{1}{2}$ times this distance, is $4\frac{3}{4}$ times $\frac{1}{9}$ of the distance to Wilbraham; what is the distance to Wilbraham?

58. Kinderhook is 16 miles from Albany, and $\frac{3}{4}$ of this distance, is $\frac{2}{3}$ times $\frac{1}{3}$ of the distance to Dalton; what is the distance to Dalton?

59. Brighton is 195 miles from Albany, and $\frac{2}{5}$ of this distance is $\frac{1}{4}$ of 2 times the distance to Worcester; what is the distance to Worcester?

60. Grafton is 162 miles from Albany, and $\frac{5}{9}$ of this distance is $\frac{1}{3}$ of 3 times the distance to Westfield, lacking 2 miles; how far is it to Westfield?

LESSON XIX.

1. $\frac{2}{3}$ of 9 is $\frac{3}{5}$ of how many times $\frac{1}{5}$ of 25?
2. $\frac{3}{4}$ of 16 is $\frac{1}{3}$ of how many times $\frac{1}{7}$ of 21?
3. $\frac{4}{5}$ of 40 is $\frac{2}{3}$ of how many times $\frac{1}{2}$ of 16?
4. $\frac{7}{8}$ of 80 is $\frac{5}{8}$ of how many times $\frac{1}{3}$ of 21?

5. $\frac{3}{4}$ of 36 is $\frac{1}{3}$ of how many times $\frac{3}{4}$ of 12?
6. $\frac{8}{9}$ of 45 is $\frac{2}{7}$ of how many times $\frac{5}{7}$ of 14?
7. $\frac{5}{6}$ of 30 is $\frac{5}{3}$ of how many times $\frac{1}{2}$ of 10?
8. $\frac{11}{12}$ of 48 is $\frac{1}{3}$ of how many times $\frac{3}{7}$ of 7?
9. $\frac{4}{5}$ of 45 is $\frac{2}{9}$ of how many times $\frac{3}{4}$ of 8?
10. $\frac{5}{7}$ of 35 is $\frac{1}{2}$ of how many times $\frac{3}{7}$ of $11\frac{2}{3}$?
11. $\frac{7}{9}$ of 54 is $\frac{2}{7}$ of how many times $\frac{2}{3}$ of $10\frac{1}{2}$?
12. $\frac{3}{5}$ of 25 is $\frac{3}{4}$ of how many times $\frac{2}{5}$ of 10?
13. $\frac{4}{7}$ of 28 is $\frac{2}{5}$ of how many times $\frac{4}{5}$ of 25?
14. $\frac{4}{9}$ of 18 is $\frac{2}{9}$ of how many times $\frac{3}{4}$ of 12?
15. $\frac{5}{6}$ of 36 is $\frac{3}{8}$ of how many times $\frac{5}{6}$ of 12?
16. $\frac{4}{9}$ of 54 is $\frac{2}{3}$ of how many times $\frac{3}{4}$ of 16?
17. $\frac{7}{8}$ of 32 is $\frac{2}{8}$ of how many times $\frac{2}{3}$ of 9?
18. $\frac{8}{9}$ of 108 is $\frac{2}{3}$ of how many times $\frac{2}{3}$ of $\frac{2}{5}$ of 15?
19. $\frac{3}{4}$ of 40 is $\frac{5}{12}$ of how many times $\frac{1}{2}$ of $\frac{4}{5}$ of 20?
20. $\frac{4}{5}$ of 20 is $\frac{2}{3}$ of how many times $\frac{2}{3}$ of $\frac{3}{4}$ of 12?

LESSON XX.

1. If 1 horse eat $\frac{1}{4}$ of a bushel of oats in 1 day, how many horses will eat a bushel in the same time?

2. If the wages of 8 weeks amount to $48, what will the wages of $2\frac{3}{4}$ weeks amount to?

3. A ship's crew of 12 men have provision for 5 months; how many months would it last 5 men?

4. A man gained $14 by selling a watch for $1\frac{2}{5}$ times what it cost him; how much did it cost him?

5. There is a pole, $\frac{7}{9}$ of its length is under water, and 9 feet out; how long is the pole?

6. A pole is standing in the water, so that 15 feet is above the water, which is $\frac{3}{7}$ of the whole length of the pole; how long is the pole?

7. $\frac{3}{4}$ is $\frac{3}{5}$ of what number?

8. If 8 horses can in 1 day eat 4 bushels of oats, how many days would it take 1 horse to eat 1 bushel?

9. If 3 horses can in 1 day eat $1\frac{4}{5}$ bushels of oats, how many bushels can 1 horse eat in 4 days?

10. If 1 horse in 2 days can eat 6 bushels of corn, how many bushels will 4 horses eat in 3 days?

11. If 4 horses eat 16 bushels of provender in 2 days, how many bushels will 3 horses eat in 12 days.

12. How many tons of hay will 3 horses consume in 4 days, if 4 horses will in $\frac{1}{2}$ of a day consume $\frac{4}{5}$ of a ton?

13. How many hundred weight of hay can 3 horses consume in 25 days, if 2 horses in $\frac{1}{4}$ of a day consumed $\frac{7}{400}$ of a hundred weight?

14. How many days would it require, for 4 men to cut 16 cords of wood, if 1 man in 1 day cut $\frac{1}{3}$ of a cord?

15. How many men will be required, to earn 20 dimes in 4 days, if 4 men in $2\frac{3}{4}$ days earn 11 dimes?

16. If it require 6 days for 2 men to lay 36 rods of wall, how many men can in $\frac{1}{2}$ of the time build 72 rods of similar wall?

17. If in 4 days 3 men accomplish a certain piece of work, how many men will be required, to perform a piece of work 4 times as large in 2 days?

18. If 4 men in 8 days perform a certain piece of work, how many men will be necessary to accomplish 3 times as much work in $\frac{3}{4}$ of a day?

19. If 1 horse eat 1 bushel of oats in 4 days, in how many days would 6 horses eat 48 bushels?

20. If 2 men in $\frac{1}{3}$ of a day earn $\frac{5}{12}$ of a dollar, how many days would it take 3 men to earn $\frac{3}{4}$ of a dollar?

21. If it require $\frac{1}{8}$ of a bushel of oats to feed 4

horses $\frac{1}{3}$ of a day, how many horses would it take to consume 9 bushels in $\frac{3}{5}$ of a day?

☞ SUGGESTION.—*Review unless the pupils thoroughly understand the preceding Chapters.* ☜

CHAPTER VII.

LESSON I.

1. \$24 is $\frac{3}{5}$ of twice as much as a cask of wine cost; what did the wine cost?

2. Bought 30 barrels of flour, and $\frac{4}{5}$ of the number of barrels, equaled $\frac{1}{5}$ as many dollars as they all cost; what did 1 barrel cost?

3. 35 is $\frac{5}{8}$ of how many times $\frac{2}{3}$ of 4?

4. A farmer, being asked how many sheep he had, answered, that 160 was $\frac{2}{5}$ of 10 times his number; now many sheep had he?

5. Mr. B, being asked the value of his horse, said, that \$54 was $\frac{6}{11}$ of 3 times its value; what was the value of his horse?

6. 72 is $\frac{8}{9}$ of how many times $\frac{3}{4}$ of 12?

7. 36 is $\frac{3}{4}$ of how many times $\frac{2}{3}$ of 12?

8. 48 is $\frac{2}{3}$ of how many times $\frac{1}{2}$ of 18?

9. 56 is $\frac{8}{9}$ of how many times $\frac{7}{8}$ of 8?

10. 60 is $\frac{3}{5}$ of how many times $\frac{5}{8}$ of 16?

11. 84 is $\frac{12}{5}$ of how many times $\frac{1}{5}$ of 25?

12. A spent \$60, which was $\frac{5}{6}$ of 4 times as much as he was worth; how much was he worth?

13. B sold 9 sheep, which was $\frac{3}{10}$ times $\frac{1}{5}$ of his whole flock; how many sheep had he remaining?

14. D, at a game of cards, lost \$20, which was 4 times $_{3}$ of all the money he had; how much had he?

15. C found \$45, which was $\frac{4}{5}$ of 3 times as much

as he already had; how much more did he find than he had at first?

16. A boy lost 9 marbles, which was $\frac{3}{8}$ of twice as many as he had at first; how many had he remaining?

17. A boy gave away 8 apples, which was $\frac{2}{7}$ of twice as many as he then had; how many had he a first?

18. 12 is $\frac{3}{5}$ times $\frac{3}{7}$ of what number?

19. 36 is $\frac{6}{11}$ times $\frac{3}{5}$ of what number?

20. Jeremiah is 18 years old, and his age is $\frac{3}{4}$ times $\frac{2}{3}$ of his father's age; how old is his father?

21. Mary gave 6 cents for a comb, which was $\frac{2}{5}$ times $\frac{1}{2}$ of all the cents she had; how many cents had she?

22. Martha gave 8 cents for a pine-apple, which was $\frac{2}{7}$ times $\frac{2}{5}$ of all her money; how many apples could she have bought with the money she had remaining, at 2 cents apiece?

23. Henry had 20 marbles, which was $\frac{2}{3}$ of twice as many as Harry had; how many had Harry?

24. Margaret is 16 years old, and her age is $\frac{8}{9}$ of 3 times Martha's age; how old is Martha?

25. $\frac{3}{4}$ is $\frac{2}{3}$ of twice as much as what number?

26. A man bought a horse for $60, which was $\frac{3}{7}$ of twice as much as he sold him for; how much did he gain by tho bargain?

27. A horse was sold for $40, which was $\frac{4}{5}$ times $\frac{5}{6}$ of what he was worth; what was the value of the horse?

28. A man, when he was married, was 20 years of age, which was $\frac{4}{5}$ times $\frac{5}{3}$ of the age of his wife; how old was she?

29. Shepherd was worth $160, which was $\frac{3}{7}$ times $\frac{1}{10}$ of his father's fortune; what was the father's fortune?

30. A and B were playing cards, B lost $14 which

was $\frac{7}{10}$ times $\frac{2}{3}$ as much as A then had; and when they commenced $\frac{5}{8}$ of A's money equaled $\frac{3}{7}$ of B's. How much had each when they began to play?

LESSON II.

1. A boy, after spending $\frac{3}{5}$ of all his money, found, that 16 cents was all he had remaining; how much had he at first?

SOLUTION.—Let $\frac{5}{5}$ equal all he had at first. Then after spending $\frac{3}{5}$ of it, he must have had remaining $\frac{5}{5}-\frac{3}{5}$, which is $\frac{2}{5}$. And this by the condition of the question is 16 cents. If 16 cents is $\frac{2}{5}$ of what he had at first, $\frac{1}{5}$ of what he had, is $\frac{1}{2}$ of 16 cents, which is 8 cents; and $\frac{5}{5}$ (what he had at first) is 5 times 8 cents, which are 40 cents.

2. Ruth, after losing $\frac{2}{3}$ of all her roses, had only 3 remaining; how many had she at first?

3. Jane gave $\frac{3}{5}$ of all her flowers to Ann, and had 4 remaining; how many did she give to Ann?

4. George, after eating $\frac{9}{13}$ of all his oranges, found that 8 oranges were all he had remaining; how many had he at first?

5. A boy expended $\frac{1}{9}$ of his money for a pie, $\frac{2}{9}$ for a ball, $\frac{3}{9}$ for a top, and had 6 cents remaining; how many cents had he at first?

6. In a certain school $\frac{1}{2}$ of the scholars study grammar, $\frac{1}{3}$ study arithmetic, and the remainder, which is 10, study geography; how many scholars in all, and how many attending to each study?

7. A third part of an army was killed, $\frac{1}{4}$ part taken prisoners, and 300 remained unhurt; how many were there in the army?

8. If from my age you subtract $\frac{1}{2}$ and $\frac{2}{5}$ of my age, the remainder would be 2 years; how old am I?

9. B, being asked how many pigeons he caught, said, that if to $\frac{4}{5}$ of the number, 36 were added, the

sum would equal twice the number. How many did he catch?

10. If to $\frac{3}{4}$ of the cost of B's horse you add \$100, the sum will be twice the cost of the horse; what was the cost of the horse?

11. A gentleman, after spending $\frac{3}{7}$ of his fortune and $\frac{1}{2}$ of the remainder, had \$2400 remaining; what was his fortune?

12. A gambler lost $\frac{3}{4}$ of all his money, and the next night he won $\frac{2}{3}$ as much as he lost the night before; he then had \$90: how much had he at first?

13. A man willed $\frac{2}{9}$ of his fortune to his wife, $\frac{3}{8}$ to his son, $\frac{1}{3}$ to his daughter, and the remaining \$200 he bestowed for charitable purposes. What was his fortune, and what did he will to each?

14. A traveler had stolen from him $\frac{5}{7}$ of all his money; the thief was caught, but not until he had spent $\frac{1}{3}$ of it, the remainder (\$100) was given back; how much had he at first?

15. If to $\frac{1}{2}$ of the cost of A's watch you add \$10, the sum will be \$21; what was the cost of his watch?

16. If to $\frac{3}{5}$ of B's age you add 7 years, the sum would be 39 years; how old is B?

17. A drover, being asked how many sheep he had, said, if to $\frac{1}{3}$ of my flock you add the number $9\frac{1}{2}$, the sum will be $99\frac{1}{2}$; how many sheep had he?

18. $\frac{4}{7}$ of the length of a pole is in the water, and 12 feet in the air; how long is the pole?

19. If to $\frac{7}{8}$ of A's age you add 16 years, the sum will be $1\frac{1}{8}$ times his age; how old is he?

20. A man, being asked how many pigeons he caught, replied, that if to $\frac{3}{4}$ of the number he caught you add 20, the sum lacking 4 would be $1\frac{1}{2}$ times the number; how many did he catch?

LESSON III

1. Divide the number 36 into two parts, which shall be to each other as 7 to 2.

SOLUTION.—Since the two parts are to be to each other as 7 to 2, we must divide 36 into 7 + 2, which are 9 equal parts; and 7 of the parts will be one of the numbers, and 2 of them the other. $\frac{1}{9}$ of 36 is 4, and $\frac{7}{9}$ is 7 times 4, which are 28 (the first number,) and $\frac{2}{9}$ is 2 times 4, which are 8 (the other number.)

2. Two men hired a pasture for $72; one put in 7 horses, and the other 2 horses; what ought each to pay?

3. A and B hired a pasture for $14; A put in 4 cows, and B put in 3 cows; what ought each to pay?

4. A and B bought a lottery ticket for $5; A paid $3, and B paid $2. They drew a prize of $60; what was each one's share?

5. Two men bought 40 mules; the first paid $5 as often as the other $3. How many mules ought each to receive?

6. Mary and Elizabeth went to school 80 days, and as often as Mary went 3 days, Elizabeth went 5 days; how many days did each go?

7. Ruben had 7 cents, and Blake 4 cents; they paid all their money for 22 apples; how many ought each to receive?

8. Three men bought a lottery ticket for $12; the first paid $2, the second $7, and the third $3. They drew a prize of $240; what was each man's share?

9. Three men hired a pasture for $24; the first put in 2 horses, the second put in 3 horses, and the 3d put in 4 horses; what ought each to pay?

10. A man, failing in business, was able to pay only $\frac{4}{5}$ of his debts; what will that man receive to whom he owes $90?

11. A man, meeting an equal number of poor wo-

men and boys, gave to each woman 7 dimes, and to each boy 2 dimes; and to them all he gave \$9: how many women and boys were there respectively?

12. Two men bought a barrel of fish for \$9; the first paid \$4, the second \$5; what part of the barrel belongs to each?

13. A farmer gave 35 bushels of rye to 2 of his poor neighbors; to the first he gave 1 bushel as often as to the other $\frac{3}{4}$ of a bushel: how many bushels did each receive?

14. Three men hired a pasture for \$36; the first put in 3 horses, the second 2 horses, and the third 4 horses; how much ought each to pay?

15. Two men hired a pasture for \$60; the first put in 4 horses for 2 weeks, and the second put in 3 horses for 4 weeks. What ought each to pay?

16. Three men hired a pasture for \$15; the first put in 4 sheep for 5 weeks, the second put in 8 sheep for 5 weeks, and the third put in 10 sheep for 9 weeks. What ought each to pay?

17. Two men entered into partnership; the first put in \$40 for 10 months and the second put in \$80 for 5 months; they gained \$95: what was each man's share of the gain?

18. A and B agreed to cut a field of wheat for \$20; A sent 5 men for 4 days, and B sent 3 men for 10 days; what ought A and B to receive respectively?

19. Divide \$56 between A and B, giving to A \$1 as often as $\frac{2}{5}$ of a dollar to B.

20. A and B hired a pasture for \$24; A put in 4 sheep for 10 weeks, and B put in 2 horses for 10 weeks; what ought each to pay, providing 2 sheep in week eat as much as a horse in the same time?

LESSON IV.

1. B had 4 apples more than A, and they together had 14; how many had each?

SOLUTION.—By a condition of the question 4 apples + A's number equals B's number; and A's number added to B's is twice A's, + 4 apples, which equals 14 apples. Hence 14 — 4 (which is 10 apples) is twice A's number, &c.

2. Heman has 6 books more than Handford, and they together have 26; how many has each?

3. Robert has 7 marbles more than Richard, and they together have 35; how many has each?

4. Mary has 4 roses more than Martha, and they together have 24; how many has each?

5. Alice has 7 pins more than Abner, and they together have 29; how many has each?

6. $\frac{2}{5}$ of $\frac{2}{3}$ is $\frac{4}{5}$ of what number?

7. The sum of two numbers is 36, and their difference is 16; what are the two numbers?

8. A boy bought $\frac{5}{7}$ of a melon for $8\frac{1}{3}$ cents; how much is that apiece?

9. Homer and Hannah each bought an equal number of peaches; on their way home Hannah had 4 more given to her, then they together had 24: how many did each buy?

10. Two boys had each an equal number of blocks; one lost 4 of his; and they together then had only 12 remaining: how many had each at first?

11. A wagon was sold for $\$17\frac{2}{3}$, which was $\frac{2}{3}$ as much as it cost; what did it cost?

12. Hiram had twice as many strawberries as Eugene, and they together had 18 pints; how many had each?

13. Joshua had 6 cents more than twice as many as Jordon, and they together had 36; how many had each?

14. Susan had $\frac{1}{2}$ as many cents as Sarah; Sarah

lost 10 of hers; then they together had 50;—how many had each at first?

15. Thomas was returning from market with twice as many eggs as Timothy; Thomas broke 4 of his, and Timothy, 6 of his; they then together had only 50 eggs remaining. How many had each at first?

16. $\frac{3}{5}$ of a number $+ 14 = 44$; what is that number?

17. A boy, being asked his age, replied, that 3 times his age — 7 years was 23 years. How old was he?

18. A, being asked how much money he had, replied, that twice what he had + \$60, was four times \$400; how much had he?

19. Two boys have 49 marbles, but the first has 7 the most; how many has each?

20. A man bought a sheep, a cow, and a horse for \$70; the cow cost \$10 more than the sheep, and the horse cost \$20 more than the cow. What was the cost of each?

21. A man bought a melon for $18\frac{3}{4}$ cents, which was only $\frac{3}{5}$ of as much as his dinner cost; what was the cost of his dinner?

22. A gentleman bought a watch and chain for \$80; the chain cost $\frac{1}{3}$ as much as the watch; what was the cost of each?

23. A farmer bought a plow, a harness, and a horse for \$58; for the harness he gave \$6 more than for the plow, and for the horse \$34 more than for the harness. What did he give for each?

24. A boy bought twice as many oranges as lemons, and on his way home ate 4 oranges and gave 6 away; and was surprised to find that he had only 14 oranges remaining. How many of each kind did he buy?

25. 5 times a certain number — 12 is 48; what is that number?

26. $\frac{3}{4}$ of a certain number — 5 is 40; what is that number?

27. A boy, being asked his age, replied, that 11 years were 7 years more than $\frac{2}{5}$ of his age; how old was he?

28. A boy, being asked how many sheep his father had, replied, that 40 were 5 less than $\frac{3}{4}$ of his number. How many had he?

29. A boy bought 18 lemons;—for $\frac{2}{3}$ of them he paid 3 cents for 2, and for the remainder he paid 2 cents apiece; for what must he sell them apiece to gain 10 cents on the whole?

30. James, John, and Joseph together have 96 peaches; James has 2 more than John, and Joseph has as many as James and John both: how many has each?

LESSON V.

1. If a man can do a certain piece of work in 12 days, what part of it can he do in 1 day?

2. If a man can drink a barrel of beer in 20 weeks, what part of it can he drink in 1 week?

3. If it require 9 hours to empty a vessel, what part of it can be emptied in 1 hour?

4. If a family consume a barrel of pork in 30 days, what part of a barrel do they daily consume?

5. If it require 19 days to perform a certain journey, what part of it can be performed in 1 day?

6. If A can do a certain piece of work in 8 days, and B could do the same in 12 days; what part of it could each do in a day?

7. If C could mow a certain field in 4 days, and D could do the same in 6 days; what part of it could each do in a day, and how much could they together do in a day?

8. If C and D can, in 1 day, mow $\frac{5}{12}$ of a field, how long would it take them to mow the whole field?

9. How many days would it take to perform a certain piece of work, if $\frac{3}{15}$ of it can be performed in 1 day?

10. If George can do a certain piece of work in 3 days, and Granvil can do the same in 6 days; what part of it can they together do in 1 day, and how long would it take them to do the whole work?

11. If James can eat a bushel of apples in 10 days, and Ruben in 12 days; how long would 1 bushel last them both?

12. A can cut a field of wheat in 12 days, and B can do the same in 20 days; how long would it take them to cut the field when they work together?

13. A merchant bought a hogshead of molasses for $20, 10 gallons of which leaked out; how must he sell the remainder a gallon to gain $6,50 on the whole?

14. $\frac{2}{3}$ of a barrel of flour cost $$4\frac{2}{3}$, what will $\frac{4}{5}$ of a barrel cost?

15. A and B can build a boat in 20 days, but with the assistance of C, they can do it in 8 days. How long would it take C to do it alone?

16. A farmer and his son can do a piece of work in 6 days; the son, alone, can do the same in 27 days. How long would it take the father alone to do the same?

17. Three pipes, A, B, and C can fill a cistern in 2 hours, A and B can fill the same in 4 hours, and A and C can fill the same in 3 hours. How long would it take each to fill it?

18. If a barrel of beer would last a man 35 days, and, the man and his son 20 days; how long would it last the son alone?

19. A box of tea, usually, lasted a man and his wife 9 months; when the man was absent it would last the wife 12 months. How long would it have lasted the man alone?

20. A, B, and C can build a boat in 20 days, A and B, in 40 days, and A and C, in 30 days. How long would it take each separately to build it?

21. Providing A could drink a barrel of beer in 24 days, and B, in 36 days; how long would it take them together to empty a barrel, after $\frac{1}{3}$ of it had leaked out?

22. A market-woman bought 30 oranges, and had $\frac{1}{3}$ of them stolen; the remainder she sold at 3 cents each and thereby gained $\frac{2}{3}$ of a cent on each orange bought. What did they cost apiece?

23. A can do a certain piece of work in 8 days, and A and B together can do the same, in 5 days. After A did $\frac{1}{3}$ of the work, B did the remainder; how long did it take him?

24. If A can do a certain piece of work in $\frac{2}{6}$ of a day, how much can he do in 1 day?

25. If a man can chop a cord of wood in $\frac{3}{7}$ of a day, how much can he chop in 1 day?

26. Isaac can make a pair of boots in $\frac{2}{3}$ of a day, and Ira, in $\frac{2}{5}$ of a day; how many pair can they both make in 1 day?

27. Samuel can cut a cord of wood in $\frac{3}{4}$ of a day, and Theodore, in $\frac{2}{5}$ of a day; how long would it take them to cut a cord, when they work together?

28. If $\frac{3}{5}$ of an apple cost $\frac{6}{7}$ of a cent, what will $\frac{5}{8}$ of an apple cost?

29. A can mow 1 acre of grass in $\frac{2}{3}$ of a day, B, in $\frac{3}{4}$ of a day, and C, in $\frac{4}{5}$ of a day. How much more can A and B mow in a day than C?

30. If a wolf can eat a sheep in $\frac{7}{8}$ of an hour, and a bear can eat it in $\frac{3}{4}$ of an hour, how long would it take them together to eat what remained of a sheep, after the wolf had been eating $\frac{1}{2}$ of an hour?

LESSON VI.

1. Lewis, meeting some beggars, gave each of them 2 cents, and had 12 cents remaining; if he had given them 4 cents each, it would have taken all the money he had. How many beggars were there?

SOLUTION.—By the last condition of the question, he gave each beggar 2 cents more than by the first, and to them all, 12 cents *more* than by the first condition. Therefore there must have been as many beggars as 2 is contained in 12, which are 6 beggars.

2. A boy gave to each of his playmates 3 cents, and had 24 cents remaining; if he had given them each 7 cents, it would have taken all the money he had. How many playmates had he?

3. Mary gave each of her playmates 5 apples; if she had given them each 7 apples, it would have taken 12 apples more. How many playmates had she?

4. A certain number of persons gave me 10 cents each; had they given me 12 cents each, it would have amounted to 20 cents more. How many persons were there?

5. $\frac{3}{5}$ of 100 is $\frac{3}{250}$ of $\frac{1}{5}$ times the yearly salary of the President of the United States. What is his salary?

6. $40 is $\frac{2}{5}$ times $\frac{1}{100}$ of the salary yearly received by the Vice President of the United States. What is his salary?

7. Divide 35 oranges between James and Joseph, so that James may have 15 more than Joseph.

8. A cask of wine was sold for $96, which was $\frac{3}{4}$ as much again as it cost; what did it cost?

9. A quantity of cotton was sold for $560, and thereby gained $\frac{3}{4}$ of what it cost; what did it cost?

10. A and B are 187 miles apart, and are traveling towards each other; one at the rate of 8 miles an hour, and the other, 9 miles an hour; how many hours before they will meet?

11. Agnes gave 2 dimes a yard for a piece of calico; had she given 3 dimes a yard, it would have cost her 20 dimes more how many yards did the piece contain ?

12 A was ordered to buy a certain number of ranges; if he bought those at 2 cents each, he would nave had no money left,—had he bought those at 3 cents each, he would have wanted 10 cents more to have paid for them. How many oranges was he required to buy ?

13. A lady wished to buy a certain number of yards of muslin ; there were two kinds, some at 9 cents a yard, and some at 12 cents a yard. Had she taken that at 12 cents a yard, it would have cost 36 cents more than the other kind. How many yards did she wish to buy ?

14. A boy, being sent to market to buy a certain number of pounds of meat, found, if he bought beef at 5 cents a pound, he would have 39 cents remaining, but if he bought pork, at 8 cents a pound, he would have only 6 cents remaining. For how much meat was he sent ?

15. If 8 times a certain number, is 36 more than 5 times the same number ; what is that number ?

16. A boy, being asked his age, said, that 4 times his age was 24 years more than 2 times his age ; how old was he ?

17. A boy, being asked how many sheets of paper he had, said, that 4 times the number, was 18 less than 7 times the number ; how many sheets of paper had he ?

18. A person, wishing to buy some butter, found, if he bought that which was 10 cents a pound, he would have 20 cents remaining ; but if he bought that which was 12 cents a pound, he would lack 14 cents of having money enough to pay for it. How many pounds did he wish to buy ?

19. A farmer, wishing to buy a certain number of

sheep, found if he gave $2 a head, he would have $20 remaining; but if he gave $5 a head he would lack $40 of having money enough to pay for them. How many sheep did he wish to buy?

20. A, B, and C, talking of their ages; says A to B, am 4 times as old as you; says B to C, I am ½ as old as you, but says A to C, I am 40 years older than you. Required the age of each.

LESSON VII.

1. A laborer agreed to work for 40 days upon this condition; that for every day he worked, he should receive $2, and for every day he was idle, he should pay $1 for his board. At the expiration of the time, he received $50 How many days did he work?

SOLUTION.—If he had labored the whole time, he would have received 40 times $2 which are $80. But he received only $50: he, therefore, lost by his idleness $80—$50, which is $30. For every day he was idle he lost $2 (his daily wages)+$1 (the cost of his board) which are $3. If in 1 day he lose $3; he will lose $1 in ⅓ of a day, and $30 in 30 times ⅓ of a day, which are 10 days. Therefore he was idle 10 days, and worked 40—10 days, which is 30 days.

2. A man agreed to work 60 days, on this condition; that for every day he worked, he should receive $1½, and for every day he was idle, he should pay $½ for his board. At the expiration of the time, he received $68. How many days did he work?

3. A man was hired for 80 days, on this condition; that for every day he worked, he should receive 6 dimes, and for every day he was idle, he should forfeit 4 dimes. At the expiration of the time, he received $40. How many days did he work?

4. How many times ⅔ of 12 is ⅞ of 48?

5. A and B bought a quantity of flour for $50; A paid $1 as often as B ⅔ of a dollar; what part of the flour belongs to each?

6. A, B, and C built a house, which cost $500, of which B paid $100 more than A, and C paid as much as A and B both; how much did each pay?

7. A merchant sold a quantity of cloth for $84, and thereby lost $\frac{2}{5}$ of what it cost; what did it cost?

8. $7\frac{1}{2}$ is $2\frac{1}{2}$ times $\frac{2}{3}$ of what number?

9. A farmer having in his employ an equal number of men and boys; to each boy he gave $4, to each man $8; and to them all he gave $84: how many men were there?

10. Two men hired a pasture for $35; one put in 3 cows, and the other put in 4; how much ought each to pay?

11. A man sold an equal number of ducks and turkeys for 20 dimes; the ducks, at 2 dimes each, and the turkeys, at 3 dimes each: how many did he sell in all?

12. A farmer sold an equal number of ducks and turkeys; the ducks, at 4 dimes each, the turkeys, at 7 dimes each; and for the turkeys he received $3 more than for the ducks: how many of each kind did he sell?

13. There are two baskets, which contain 37 apples; one of which contains 17 the most: how many apples are there in each?

14. Charles and Henry together have 49 marbles, and Charles has 7 more than twice as many as Henry: how many has each?

15. Philip has 20 apples more than Philo; and they together have 92: how many has each?

16. Three boys have 47 lemons; the first has 3 more than the second, and the second has 7 more than the third: how many has each?

17. A boy was hired for 20 days, on this condition; that for every day he labored, he should receive 3 dimes, and for every day he was idle, he should pay 2 dimes for his board. At the expiration

of the time he received only $1. How many days was he idle?

18. A boy bought a whistle, a whip, and a drum for 70 cents. For the whip he gave twice as much as for the drum, and for the drum, twice as much as for the whistle; what did he give for each?

19. The sum of three numbers is 54. The first is twice, and the third, 3 times the second; what are those numbers?

20. Sarah's age is $\frac{3}{5}$ of Susan's, and the sum of their ages is 24; what is the age of each?

21. $\frac{5}{6}$ of an army was killed, $\frac{2}{3}$ of the remainder taken prisoners, and 400 escaped. How many were there in the whole army?

LESSON VIII.

1. A fishing rod, the length of which was 14 feet, was broken into two pieces. The shorter piece was $\frac{3}{4}$ of the length of the longer. What was the length of each piece?

SOLUTION.—The length of the shorter, which is $\frac{3}{4}$ of the longer, added to $\frac{4}{4}$ (the length of the longer) is $\frac{7}{4}$ of the length of the longer, which is the length of both, which is 14 feet. If $\frac{7}{4}$ of the longer is 14 feet, $\frac{1}{4}$ is $\frac{1}{7}$ of 14 feet, which is 2 feet, $\frac{4}{4}$ (the length of the longer) is 1 times 2 feet, which are 8 feet, &c.

2. A pole the length of which is 20 feet, is in the air and water $\frac{3}{3}$ of the length in the air, equals the length in the water; what is the length in the air and water respectively?

3. If in 2 days a man traveled 160 miles, and $\frac{3}{5}$ of the distance he traveled the first day, equals the distance he traveled the second day; how far did he travel each day?

4. B and C together have 40 marbles; how many has each, providing $\frac{3}{5}$ of B's number is equal to C's number ?

5. From New York City to Redhook is 100 miles, and $\frac{1}{9}$ of the distance from New York to Rhinebeck, equals the distance from there to Redhook. How far from Rhinebeck to New York, and how far from Rhinebeck to Redhook ?

6. If a horse and a colt were worth $90, and the horse was worth $1\frac{1}{2}$ times as much as the colt; what was the value of each ?

7. A boy paid 70 cents for a slate and a book; how much did he pay for each, providing the book cost $1\frac{1}{3}$ times as much as the slate ?

8. If a traveler pay $1,20 for his breakfast and dinner; what did he pay for each, providing his dinner cost $\frac{5}{7}$ as much as his breakfast ?

9. A pole, the length of which is 57 feet, is in the air and water; $\frac{2}{3}$ of what is in the air + 7 feet equals what is in the water. Required the length in the air, and in the water.

10. Divide the number 108 into two such parts, that $\frac{3}{7}$ of the first + 8 shall equal the second.

11. Divide the number 97 into two such parts, that $\frac{3}{7}$ of the first + 7 shall equal the second.

12. There is a fish the length of which is 18 feet; its tail is 4 feet, and $\frac{2}{5}$ of the length of the body, equals the length of the head. What is the length of the head and body respectively ?

13. There is a fish the weight of which is 11 pounds, and $\frac{1}{2}$ of the weight of the head + 8 pounds equals the weight of the body; what is the weight of each ?

14. A ship-mast 51 feet in length, in a storm, was broken off, and $\frac{2}{3}$ of what was broken off, equaled $\frac{3}{4}$ of what remained; how much was broken off, and how much remained ?

15. A boy, being asked how many apples and

oranges he had, answered, that in all he had 36, and, that $\frac{2}{5}$ of the number of apples, equaled $\frac{1}{4}$ of the number of oranges; how many of each kind had he?

16. $\frac{2}{3}$ of one number equals $\frac{3}{5}$ of another, and their sum is 57; what are the two numbers?

17. A farmer has 290 sheep in two different fields; and $\frac{3}{4}$ of the number in the first field, equals $\frac{2}{7}$ of the number in the second; how many are there in each field?

18. A market woman was requested to buy 33 fowls, consisting of two different kinds; $\frac{1}{4}$ of the number of the first kind, was to equal $\frac{2}{3}$ of the second kind; how many of each kind must she buy?

19. A person, being asked the time of day, said, the time past noon is $\frac{1}{7}$ of the time past midnight; what was the hour?

REMARK.—Since the time past noon is $\frac{1}{7}$ of the time past midnight, the time from midnight to noon, which is 12 hours, must be $\frac{6}{7}$ of the time past midnight.

20. A person, being asked the hour of the day, said, the time past noon, is $\frac{1}{3}$ of the time past midnight; what was the hour?

21. A person, being asked the hour of the day, said, the time past noon, is $\frac{1}{3}$ of the time from now to midnight; what is the hour?

SOLUTION.—From that time to midnight is $\frac{3}{3}$, and $\frac{1}{3}$ added (the time past noon) is $\frac{4}{3}$. Consequently from noon to midnight (which is 12 hours) is $\frac{4}{3}$ of the time it lacked of being midnight, and $\frac{1}{3}$ of the time, is $\frac{1}{4}$ of 12, which is 3 hours, the time past noon.

22. What is the time of day, providing $\frac{3}{4}$ of the time from now to midnight equals the time past noon?

23. A man, being asked the hour of the day, said, that $\frac{2}{3}$ of the time past noon, equaled $\frac{2}{9}$ of the time from now to midnight; what was the time?

24. A pole the length of which was 68 feet was in the air and water; $\frac{3}{4}$ of the length in the air, equaled $\frac{2}{3}$ of

the length in the water. What was the length in the air, and in the water respectively?

25. The sum of two numbers is 176, and $\frac{3}{4}$ of the first + 4 equals $\frac{2}{3}$ of the second; what are those numbers?

26. A person, being asked the time of day, said, that $\frac{3}{5}$ of the time past midnight equals $\frac{3}{10}$ of the time from now to midnight again; what was the time?

27. Providing the time past 10 o'clock equals $\frac{3}{4}$ of the time to midnight; what o'clock is it?

28. Says A to B, $\frac{3}{4}$ of my age + 4 years equals $\frac{2}{3}$ of yours, and the sum of our ages is 74 years. What is each of their ages?

29. A person, being asked the hour of the day, replied, that $\frac{2}{3}$ of the time past noon, equaled $\frac{2}{9}$ of the time from then to midnight + $2\frac{2}{3}$ hours; what was the time?

30. A pole the length of which is 78 feet, is in the air and water; $\frac{9}{8}$ of what is in the air + 12 feet, equals $1\frac{1}{2}$ times the length in the water. What is the length in the air, and water respectively?

LESSON IX.

1. There is a fish the head of which is 4 inches long, and whose tail is as long as its head + $\frac{1}{2}$ of its body, and whose body is as long as its head and tail both; what is the length of the fish?

SOLUTION.—By a condition of the question, $\frac{1}{2}$ of the ength of body + 4 inches, is the length of the tail; and 4 inches more added, (the length of the head,) equals the length of the body, which is $\frac{2}{2}$. And if $\frac{2}{2}$, (the length of the body,) equals $\frac{1}{2}$ of the length of the body + 8 inches; the 8 inches must be $\frac{2}{2} - \frac{1}{2}$, which is $\frac{1}{2}$ of the length of the body, &c.

2. The head of a fish is 6 inches long, the tail is as

long as the head + $\frac{1}{2}$ of the body, and the body is as long as the head and tail together; what is the length of the fish?

3. The head of a fish is 12 inches long, the tail is as long as the head + $\frac{1}{2}$ of the body, and the body is as long as the head and tail both; what is the length of the fish?

4. The head of a fish weighs 10 pounds, the tail weighs as much as the head + $\frac{2}{3}$ as much as the body; and the body weighs as much as the head and tail both; what is the weight of the tail?

5. $\frac{4}{7}$ of a certain number equals $\frac{2}{7}$ of the same number + 10; what is that number?

6. A boy, being asked his age, replied, that $\frac{2}{3}$ of his age, exceeded $\frac{2}{5}$ of his age by 4 years; how old was he?

7. James, being asked how many arithmetical questions he had answered properly during the week, replied, that $\frac{3}{4}$ of the number, was 3 more than $\frac{3}{5}$ of the number; how many questions had he answered?

8. A farmer, after selling $\frac{2}{3}$ of $1\frac{1}{5}$ times as much grain as he had, had 80 bushels remaining; how much had he at first?

9. An individual, after spending $\frac{2}{3}$ of all his money, and $\frac{2}{3}$ of what then remained, had only $\$12\frac{2}{3}$ remaining; how much had he at first?

10. If $\frac{6}{7}$ of a ship be worth $\frac{4}{5}$ of her cargo, valued at 300 eagles; what is the value of the ship?

11. Dick, being asked how much money he had, said, its $\frac{1}{2}$ exceeded its $\frac{3}{8}$ by \$2; how much had he?

12. A tree by falling, was broken into three pieces; the top part was 10 feet long, the bottom part was as long as the top + $\frac{3}{5}$ of the middle, and the middle part was as long as the other two: what was the length of the tree, and of each piece?

13. A man bought a hat, a coat, and a watch; the hat cost \$6, the watch cost as much as the hat + $\frac{2}{3}$ of the

cost of the coat, and the coat cost as much as the hat and watch both; what was the cost of each, and of all?

14. $\frac{2}{3}$ of $\frac{3}{4}$ is how many times $\frac{2}{5}$?

15. If $\frac{3}{4}$ of a ton of hay cost $\frac{3}{5}$ of an eagle, how many dollars will $\frac{1}{2}$ of a ton cost?

16. A third and $\frac{1}{2}$ of a third of 10 is $\frac{5}{8}$ of what number?

17. If from a certain number you take its $\frac{1}{2}$ and its $\frac{1}{3}$, the remainder will be $13\frac{2}{3}$, what is that number?

18. After spending $\frac{2}{3}$ of my money, I earned $\frac{2}{3}$ as much as I spent, and then had only $20 less than what I had at first; how much had I at first?

19. The head of a fish is 8 inches long, the tail is as long as the head and $\frac{1}{2}$ of the body + 10 inches, and the body is as long as the head and tail both; what is the length of the fish?

20. The head of a fish is 12 inches long; its tail is 10 inches longer than its head increased by $\frac{1}{2}$ the length of the body; and its body is 20 inches longer than its head and tail together. What is the length of the fish?

LESSON X.

1. James is 20 years old, and John is 4 years old; in how many years will James, who is now 5 times as old as John, be only twice as old?

REMARK.—Four years ago James was 16, and in 16 years more he will be twice as old as John.

2. Sarah is 10 years old, and Sally is 4; in how many years will Sally be $\frac{1}{2}$ as old as Sarah?

3. Jacob is 40 years old, and Alfred is 2; in how many years will Alfred be $\frac{1}{2}$ as old as Jacob?

4. If the third of 6 be 3, what will the fourth of 20 be?

5. If $\frac{2}{3}$ of 12 be 10, what will $\frac{2}{5}$ of 10 be?

6. What is a third and a half of $1\frac{1}{2}$ times 10?

7. Divide the number 85 into two parts, that shall be to each other as $\frac{2}{3}$ to $\frac{3}{4}$.

8. When A was married, he was 3 times as old as his wife, but when they had been married 15 years, he was only twice as old as she was; how old was each when they were married?

REMARK.—The conditions of the above question give the following:

9. Three times a certain number + 15 equals twice the same number + 30; what is that number, and what is 3 times the same number?

10. Once times a certain number + 15 equals $\frac{2}{3}$ of the same number + 30; what is that number, and what is $\frac{1}{3}$ of the same number?

11. When I first met Mr. A, I was $\frac{1}{2}$ as old as he was, and in 12 years after that, I was $\frac{3}{4}$ as old as he was; what was each of our ages when we first met?

12. There are two numbers, one of which is 4 times the other; but if to each 20 were added, one will be double the other: what are these numbers?

13. When B was married, he was 3 times as old as his wife; but after they had been married 60 years, $\frac{2}{3}$ of his age equaled hers; what was the age of each when they were married?

14. A hound takes 3 leaps to a fox 4, and 3 of the hound's leaps are equal to 6 of the fox's; how many leaps must the hound take to gain 1 on the fox?

15. If the hound takes $1\frac{1}{2}$ leaps to gain 1, on the fox; how many must he take to gain 20 on the fox?

16. A hare is 20 leaps before a hound, and takes 4 leaps to the hound 3; and 3 of the hound's leaps are equal to 6 of the hare's. How many leaps must the hound take to catch the hare?

17. A fox is 60 leaps before a hound, and takes 5 leaps to the hound 2; and 4 of the hound's leaps equals 12 of the fox's. How many leaps must the hound take to catch the fox?

18. Alfred is 60 steps before Silas, and takes 9

steps to Silas 6; and 3 of Silas' steps equals 7 of Alfred's. How many steps, at this rate, will each take before they will be together?

REMARK.—*A box of glass contains 50 square feet, or as nearly as may be.*

19. How many panes of glass in a box, providing they are 6, by 8 inches?

REMARK.—Find the area of a pane of glass by reducing the inches to parts of a foot, and then multiply these parts together. The area of a pane 6 by 8 inches is $\frac{1}{3}$ of a square foot. The remainder may be solved as follows:

SOLUTION.—If 1 pane is $\frac{1}{3}$ of a square foot, to make $\frac{3}{3}$, (a square foot,) it will take 3 times 1 pane which are 3 panes; and to make 50 square feet, (1 box,) it will require 50 times 3 panes, which are 150 panes.

20. How many panes of glass in a box, providing they are 8, by 10 inches?

21. How many panes of glass in a box, providing they are 10, by 12 inches?

22. How many panes of glass in a box, providing they are 8, by 12 inches?

23. How many panes of glass in a box, providing they are 12, by 15 inches?

LESSON XI.

1. What number is that, to which if its $\frac{1}{2}$ be added the sum will be 15?

2. What number is that, to which if its $\frac{1}{2}$ be added the sum will be 24?

3. What number is that, to which if its $\frac{1}{3}$ be added, the sum will be 40?

4. What number is that, to which if its $\frac{1}{4}$ be added, the sum will be 30?

5. What number is that, to which if its $\frac{3}{8}$ be added, the sum will be 77?

6. How old is that man, to whose age if you add its $\frac{1}{3}$ and its $\frac{2}{5}$, the sum will be 104 years ?

7. What number is that, which being increased by its $\frac{1}{2}$, its $\frac{1}{3}$ and 18 more will be doubled ?

8. A man, being asked his age, said, that if his age were increased by its $\frac{3}{5}$ and 20 more, the sum would be double his age. What was his age ?

9. Suppose I buy a certain number of boxes of butter, at $2 a box, and as many more at $4 a box ; and sell them all, at $3 a box ; do I gain or lose, and how much ?

10. A boy, being asked how many oranges he had, replied, that if his number were increased by its $\frac{2}{3}$, its $\frac{3}{4}$, and 38 more, the sum would equal 3 times his number. How many had he ?

11. Suppose I buy a certain number of melons ; some at 10 cents each, and as many more at 40 cents each ; and sell them all, at 30 cents apiece : how much do I gain on each melon ?

12. If by selling 1 apple I lose $\frac{3}{20}$ of a cent, how many apples, at this rate, must I sell to lose 6 cents ?

13. A boy bought a certain number of lemons for 2 cents apiece, and as many more, at 4 cents apiece ; and sold them out, at the rate of 3 for 5 cents : did he gain or lose, and how much ?

14. A woman bought a certain number of apples, at the rate of 2 for a cent, as many more, at the rate of 3 for a cent ; and sold them all, at the rate of 5 for 2 cents, and by so doing, lost 4 cents. How many of each kind did she buy ?

15. A woman bought a certain number of eggs, a. the rate of 3 for a cent, and as many more, at 4 for a cent ; and sold them out, at the rate of 8 for 3 cents, and by so doing, gained 4 cents. How many eggs did she buy ?

16. Three men agreed to share $510 in the proportion of $\frac{1}{2}$, $\frac{2}{3}$ and $\frac{1}{4}$; how much must each receive ?

17. A's fortune is to B's as $\frac{1}{2}$ to $\frac{1}{3}$; and they together have $100; how much has each?

18. The difference of two numbers is 15, which is $\frac{3}{4}$ of twice as much as the smaller number; what are these two numbers?

19. A merchant bought a certain number of yards of cloth, at the rate of 2 yards for $1, and as many more, at the rate of 5 for $1; and sold them all, at the rate of 10 yards for $3; and thereby lost $8. How many yards did he buy?

20. A man bought a certain number of melons, at the rate of 4 for $1, and as many more, at the rate of 10 for $1; and sold them all, at the rate of 8 for $2, and thereby gained $6. How many melons did he buy?

LESSON XII.

1. Mary has twice as many apples as Sarah, and they together have 12; how many has each?

REMARK.—By the condition of the question Mary has 2 apples as often as Sarah 1. Consequently Mary must have $\frac{2}{3}$, and Sarah $\frac{1}{3}$ of the 12 apples.

2. Divide 18 into two such parts, that one shall be twice the other.

3. Divide 21 oranges between two boys, so that one may have twice as many as the other.

4. Franklin and Francis together have 15 quarts of nuts, but Franklin has twice as many as Francis; how many has each?

5. Robert has twice as many cents as Harry, and they together have 24; how many has each?

6. Divide the number 27 into two parts, that shall be to each other as 1 to 2.

7. Harriet is twice as old as Ellen, and the sum of their ages is 30 years; what is the age of each?

8 A and B are 36 rods apart, and travel towards

each other; how far will each travel before they meet, providing A travels twice as fast as B?

9. What number must be added to twice itself, that the sum may be 57?

10. A, after spending $\frac{1}{4}$ of all his money, and $\frac{2}{5}$ of the remainder lacking $4, had only $14 remaining; how much had he at first?

11. Divide the number 48 in two such parts, that one shall be $\frac{3}{5}$ of the other.

12. In a certain school, there are 3 times as many boys as girls, and in all there are 52 pupils; how many boys, and how many girls in the school?

13. James and Jackson together have 45 marbles, but James has only $\frac{1}{4}$ as many as Jackson; how many has each?

14. A man and his son together earned $280 in a year; how much does each earn, providing the boy earns only $\frac{1}{3}$ as much as his father?

15. A boy bought a melon and a citron for $1; how much did each cost, providing the melon cost only $\frac{1}{3}$ as much as the citron?

16. A man bought a horse and a saddle for $120; the saddle cost only $\frac{1}{9}$ as much as the horse; what was the cost of each?

17. A man, being asked the cost of his oxen, said, that his oxen and wagon together cost $240, and that the oxen cost twice as much as the wagon; what was the cost of each?

18. A man bought a sheep, a hog, and cow for $42; for the hog, he gave twice as much as for the sheep, and for the cow, 3 times as much as for the sheep. How much did he give for each?

19. A farmer and his two sons earned $560 in 1 year; the father earned twice as much as his elder son, and the elder son earned twice as much as the younger son. How much did each earn?

20. A, B, and C together, in 1 day, can dig 105

bushels of potatoes · A digs $\frac{1}{2}$ as much as B, and B, $\frac{1}{2}$ as much as C. How many bushels can each dig in a day?

21. A man bought three pieces of cloth for $160; the first piece cost only $\frac{1}{3}$ as much as the second, and the second, only $\frac{1}{4}$ as much as the third. What did each piece cost?

22. In an army consisting of 20,000 men; 3 times as many were wounded as were killed, and 4 times as many remained unhurt as were wounded. How many were killed, wounded, and unhurt respectively?

23. $\frac{1}{2}$ of A's, + $\frac{2}{3}$ of B's money equals $5500; and $\frac{2}{3}$ of B's money is 4 times $\frac{1}{2}$ of A's. What is each of their fortunes?

24. Herman and Byron together have 60 blocks; and Byron owns $\frac{2}{3}$ as many as Herman; how many has each?

25. Divide the number 60 into two parts, that shall be to each other as $\frac{1}{2}$ to $\frac{3}{4}$.

26. Adelia and Louisa are to share 14 apples in the proportion of 4 to 3; how many ought each to receive?

27. The sum of Mary's and Hezekiah's age is 24 years; how old is each, providing Hezekiah is only $\frac{2}{3}$ as old as Mary?

28. Henry and his father can thrash 35 bushels of oats in a day; how many does each thrash, if Henry thrashes only $\frac{2}{3}$ as much as his father?

29. A pole, whose length is 70 feet, is in the air and water; how much is in the air and water respectively, if $\frac{3}{4}$ of what is in the air equals what is in the water?

30. Divide the number 36 into two parts, that shall be to each other as 5 to 4.

31. Divide the number 45 into two parts, that shall be to each other as 1 to $\frac{1}{2}$.

32. A and B together own \$450; but A owns only $\frac{3}{5}$ as much as B: how much belongs to each?

33. A man died, and left \$7200 to be divided between his son and daughter, in the proportion of 1 to $\frac{5}{7}$. What ought each to receive?

34. In a mixture of tea consisting of 48 pounds, there was $\frac{1}{3}$ as much poor, as good tea: how much of each kind was there?

35. A man bought a cow and a horse for \$96; the cow cost $\frac{3}{5}$ as much as the horse: what was the cost of each?

36. Moses has only $\frac{3}{7}$ as many chestnuts as Aaron; and they together have 40 quarts: how many belongs to each?

37. Divide the number 49, into two parts, that shall be to each other as 1 to $\frac{2}{5}$.

38. A hound ran 60 rods before he caught a fox; and $\frac{2}{3}$ the distance the fox ran before he was caught, equaled the distance he was ahead when they started. How far did the fox run, and how far in advance of the hound was he, when the chase commenced?

39. The sum of two numbers is 140, and the larger is to the smaller as 1 to $\frac{5}{9}$; what are the two numbers?

40. A and B together owe \$69, but B owes only $\frac{11}{12}$ as much as A; how much does each owe?

41. Thomas and Thornton found \$240, but could not agree about the division of it; they therefore threw it on the floor and each got what he could; it so happened that Thomas got only $\frac{3}{5}$ as much as Thornton. How much did each get?

42. In a certain school consisting of 48 pupils, there are $1\frac{2}{7}$ times as many boys as girls; how many boys, and how many girls in the school?

43. A gold and a silver watch were bought for \$160; the silver watch cost only $\frac{1}{7}$ as much as the gold one; what was the cost of each?

44. Divide the number 17 into two parts, that shall be to each other as $\frac{2}{3}$ to $\frac{3}{4}$.

45. A farmer had 180 sheep in two fields, and $\frac{1}{4}$ of the number in the first field equaled $\frac{1}{5}$ of the number in the second; how many in each field?

46. Divide 88 into two parts that shall be to each other as $\frac{2}{3}$ to $\frac{4}{5}$.

47. $\frac{3}{5}$ of the distance a hare ran after a hound started after her, equaled the distance she was before the hound when they started; how far did the hare run before she was caught, providing the hound ran 80 rods to overtake her?

48. A and B started from the same point, and ran in the same direction; B ran 60 rods; then $\frac{1}{11}$ of the distance A had run equaled the distance A was ahead of him. How much did A gain on B in running 60 rods?

49. A fishing rod, the length of which is 24 feet, is in two parts; $\frac{3}{5}$ of the longer part equals the length of the shorter. How long is each part?

50. A hound ran 90 rods before he caught a deer; the deer ran 44 times as far as he was ahead of the hound when they started, before it was overtaken. How far ahead of the hound was the deer when the chase commenced?

51. $\frac{2}{3}$ of A's number of sheep + $\frac{3}{4}$ of B's number, equals 900; how many sheep has each, providing $\frac{3}{4}$ of B's number is twice $\frac{2}{3}$ of A's number?

LESSON XIII.

1. A person had two silver cups, and only one cover for both. The first cup weighed 6 oz. If the first cup be covered, it will weigh twice as much as the second, but if the second cup be covered it will weigh

3 times as much as the first. What is the weight of the second cup, and cover ?

SOLUTION.—By the last condition of the question, 3 times the weight of the first cup, (18 ounces,) equals the weight of the second cup and cover. Consequently the two cups and cover weigh 18, + 6 ounces, which are 24 ounces. And by the first condition, the first cup and cover weigh twice as much as the second cup. Therefore the 24 ounces must be divided into two parts, which are to each other as 2 to 1. One of these parts will be the weight of the second cup, and 2, the weight of the first cup and cover, &c.

2. A lady has two silver cups, and only one cover for both. The first cup weighs 8 ounces. The first cup and cover weigh 3 times as much as the second cup ; and the second cup and cover weigh 4 times as much as the first cup. What is the weight of the second cup, and cover ?

3. A man bought a hat, a coat, and a vest for $40. The hat cost $6 ; the hat and coat cost 9 times as much as the vest. What was the cost of each ?

4. A boy bought a squirrel, a rabbit, and a bird. The squirrel cost 15 cents. The squirrel and rabbit cost twice as much as the bird ; and the rabbit and bird cost 3 times as much as the squirrel. What was the cost of the bird and rabbit respectively ?

5. A farmer bought a cow, an ox, and a horse ; the cow cost $20. The cow and ox together cost 3 times as much as the horse ; the ox and horse together cost 4 times as much as the cow. What was the cost of the ox and horse respectively ?

6. A man bought two horses and a saddle. The younger horse cost $40. The saddle cost $\frac{1}{7}$ as much as both horses ; and the younger horse cost $\frac{1}{3}$ as much as the other horse and saddle together. What did the saddle and older horse cost respectively ?

7. A man traveled three successive days. The first day he traveled 30 miles, which was $\frac{1}{4}$ of the distance he traveled the other two days ; and 4 times the distance he traveled the second day equaled the dis-

tance he traveled the first and third days. How far did he travel each day ?

8. A is worth $1000, and B and C together are worth 9 times as much as A ; and C is worth $\frac{1}{9}$ as much as A and B. How much is B and C worth respectively ?

9. A's coat cost $20, and his vest and hat together cost 5 times as much as his coat ; and 3 times the cost of his vest equaled the cost of both coat and hat. What was the cost of his vest and hat respectively ?

10. B's harness cost $120, which was $\frac{1}{3}$ of the cost of his horse and sleigh ; and the harness and horse together cost twice as much as the sleigh. How much did the horse and sleigh cost respectively ?

11. A tree in falling broke into three unequal pieces. The top piece was 8 feet long, which was $\frac{1}{5}$ of the length of the other two pieces ; and 3 times the length of the bottom piece, equals the length of the other two pieces. How long was the tree, and how long was each piece ?

12. Find the ages of A, B, and C, by knowing that A is 20 years old, and that the sum of B's and C's age is 4 times A's age ; and that C's age is $\frac{1}{9}$ of the sum of A's and B's age.

13. Find the fortunes of A, B, C, D, E and F, by knowing that, A is worth $20, which is $\frac{1}{4}$ as much as B and C are worth, and that C is worth $\frac{1}{3}$ as much as A and B ; and, also that if 19 times the sum of A's, B's and C's fortune were divided in the proportion of $\frac{3}{4}$, $\frac{1}{2}$, and $\frac{1}{3}$ it would respectively give $\frac{3}{4}$ of D's, $\frac{1}{2}$ of E's and $\frac{1}{3}$ of F's fortune.

CHAPTER VIII.

LESSON I.

REMARK.—In our calculations on interest (as heretofore mentioned) we shall reckon 30 days to the month, and 12 months to the year. Although, this, mathematically speaking, will not produce a result *strictly accurate*; yet it will be sufficiently *correct*, for all practical purposes.

1. Reduce 2 years and 4 months to the fraction of a year.

SOLUTION.—4 months is what part of a year? There are 12 months in 1 year, therefore 1 month is $\frac{1}{12}$ of a year; and 4 months is 4 times $\frac{1}{12}$, which is $\frac{4}{12}$ or $\frac{1}{3}$ of a year. In $2\frac{1}{3}$ years, how many thirds? In 1 there are $\frac{3}{3}$, therefore 3 times the number of whole ones equal the number of thirds. 3 times 2 are 6, and $\frac{1}{3}$ added are $\frac{7}{3}$ years.

☞ REMARK.—*Always reduce a fraction to its lowest terms before performing any other operation with it.* ☜

Pupils may not be able, readily to discover the greatest number that will divide both *numerator* and *denominator*, without a remainder. Consequently, the most expeditious way will be, to continue to divide by the least number that is contained in both numerator and denominator without a remainder, until the fraction is reduced to its lowest terms.

2. Reduce 1 year and 3 months to the fraction of a year.

3. Reduce 3 years and 5 months to the fraction of a year.

4. Reduce 4 years and 10 months to the fraction of a year.

5. Reduce 7 years and 9 months to the fraction of a year.

6. Reduce 8 years and 8 months to the fraction of a year.

7. Reduce 12 years and 7 months to the fraction of a year.

8. Reduce 11 years and 11 months to the fraction of a year.

9. Reduce 6 years and 6 months to the fraction of a year.

10. Reduce 9 years and 8 months to the fraction of a year.

11. Reduce 2 years 4 months and 15 days to the fraction of a year.

SOLUTION.—15 days is what part of a month? There are 30 days in one month, therefore 1 day is $\frac{1}{30}$ of a month, and 15 days is 15 times $\frac{1}{30}$, which is $\frac{15}{30}$, or $\frac{1}{2}$ of a month. $4\frac{1}{2}$ months, or $\frac{9}{2}$ months, is what part part of a year? There are 12 months in one year, therefore, 1 month is $\frac{1}{12}$ of a year, and $\frac{1}{2}$ of a month is $\frac{1}{2}$ of $\frac{1}{12}$, which is $\frac{1}{24}$ of a year; and if $\frac{1}{2}$ of a month is $\frac{1}{24}$ of a year, 9 halves is 9 times $\frac{1}{24}$, which is $\frac{9}{24}$, or $\frac{3}{8}$ of a year. In $2\frac{3}{8}$ years, how many eighths?

12. Reduce 4 years 7 months and 6 days to the fraction of a year.

13. Reduce 5 years 9 months and 18 days to the fraction of a year.

14. Reduce 1 year 7 months and 18 days to the fraction of a year.

15. Reduce 2 years 7 months and 6 days to the fraction of a year.

16. Reduce 3 years 3 months and 6 days to the fraction of a year.

17. Reduce 5 years 4 months and 24 days to the fraction of a year.

18. Reduce 6 years 5 months and 18 days to the fraction of a year.

19. Reduce 7 years 11 months and 6 days to the fraction of a year.

20. Reduce 10 years 10 months and 12 days to the fraction of a year.

LESSON II.

REMARK.—*Interest* is money due for the use of money borrowed; and is estimated, at a certain rate per cent per annum, which is regulated by law.

The sum on which the interest is paid, is called the *Principal.* The sum of the *principal* and *interest* is called the *Amount.*

By 6 per cent is meant, 6 cents on 100 cents, $6 on $100, or 6 on 100 whatever be the denomination. Hence, at 6 per cent, $\frac{6}{100}$ or $\frac{3}{50}$ of the principal equals the interest.

When no time is specified 1 year is understood.

1. At 4 per cent, what part of the principal equals the interest?

SOLUTION.—If the interest on 100 cents is 4 cents, on 1 cent it is $\frac{1}{100}$ of 4 cents which is $\frac{4}{100}$ or $\frac{1}{25}$ of a cent. Therefore, at 4 per cent, $\frac{1}{25}$ of the principal equals the interest.

2. At 2 per cent, what part of the principal equals the interest?

3. At 5 per cent, what part of the principal equals the interest?

4. At 6 per cent, what part of the principal equals the interest?

5. At 8 per cent, what part of the principal equals the interest?

6. At 10 per cent, what part of the principal equals the interest?

7. At 7 per cent, what part of the principal equals the interest?

8. At 12 per cent, what part of the principal equals the interest?

9. At 18 per cent, what part of the principal equals the interest?

10. At 15 per cent, what part of the principal equals the interest?

11. At 25 per cent, what part of the principal equals the interest?

12. At 20 per cent, what part of the principal equals the interest?

13. At 24 per cent, what part of the principal equals the interest?

14. At 30 per cent, what part of the principal equals the interest?

15. At 32 per cent, what part of the principal equals the interest?

16. A man sold a quantity of goods, and thereby gained 75 per cent on the cost; what part of the cost equals the gain?

17. A man by selling a horse, gained 25 per cent on the cost; what part of the cost equaled the gain?

18. A merchant, by selling a quantity of goods, gained 85 per cent; what part of the cost equaled the gain?

19. By selling a watch, a man lost 15 per cent on the cost; what part of the cost equaled the loss?

20. If by selling a quantity of candies, a boy gains 120 per cent on the cost, what part of the cost equals the gain?

LESSON III.

☞ REMARK.—When we say $\frac{1}{20}$ of the *principal* equals the *interest*, it will be understood, that, a twentieth part of the principal, *whatever* be the *denomination* will be the interest. Thus, if $\frac{1}{20}$ of the principal equals the interest;—the interest on $1 is $\frac{1}{20}$ of a dollar; or on 1 cent, $\frac{1}{20}$ of a cent, and on 100 cents 100 times $\frac{1}{20}$, which is $\frac{100}{20}$, or 5 cents; that is 5 cents on 100 cents, or 5 *per cent.* ☜

1. If $\frac{3}{50}$ of the principal equals the interest, what is the rate per cent?

SOLUTION.—If on 1 cent the interest is $\frac{3}{50}$ of a cent, on 100 cents it is 100 times $\frac{3}{50}$ of a cent, which is $\frac{300}{50}$, or 6 per cent.

2. If $\frac{1}{50}$ of the principal equals the interest, what is the rate per cent?

3. If $\frac{2}{25}$ of the principal equals the interest, what is the rate per cent?

4. If $\frac{3}{50}$ of the principal equals the interest, what is the rate per cent?

5. If $\frac{3}{25}$ of the principal equals the interest, what is the rate per cent?

6. If $\frac{1}{25}$ of the principal equals the interest, what is the rate per cent?

7. If $\frac{1}{15}$ of the principal equals the interest, what is the rate per cent?

8. If $\frac{1}{11}$ of the principal equals the interest, what is the rate per cent?

9. If $\frac{1}{12}$ of the principal equals the interest, what is the rate per cent?

10. If $\frac{1}{8}$ of the principal equals the interest, what is the rate per cent?

11. B, by selling a quantity of goods, lost $\frac{9}{100}$ of what it cost; what did he lose per cent?

12. A woman, by selling a quantity of oranges, gained $\frac{3}{25}$ as much as they cost her; how much did she gain per cent?

13. If a merchant, by selling books, gain $\frac{7}{50}$ of what they cost; how much does he gain per cent?

14. A man, by selling a package of books, gained $\frac{3}{5}$ of what they cost; what was his gain per cent?

15. B, by selling a hogshead of molasses, gained $\frac{13}{50}$ as many dollars as it cost him; what was his gain per cent?

16. If by laying out $37, I gain a sum equal to $\frac{3}{5}$ of it, what do I gain per cent?

17. A stationer received for a quantity of books, $\frac{5}{4}$ as much as they cost him; how much did he gain per cent?

18. James received for his horse $\frac{7}{5}$ of what it cost him; how much did he gain per cent?

19. A man sold a barrel of pork for $\frac{71}{50}$ of what it cost him; how much did he gain per cent?

20. The interest on a certain principal for 4 years, was $\frac{12}{25}$ of the principal; what was the rate per cent?

Remark.—If in 4 years $\frac{12}{25}$ of the principal equals the interest, for 1 year the interest will be $\frac{1}{4}$ of $\frac{12}{25}$, which is $\frac{3}{25}$ of the principal, &c.

21. A man, being asked what per cent he received on the money he had on interest, replied, that in 10 years the interest would equal $\frac{1}{2}$ of the principal. What rate per cent did he receive?

22. $\frac{4}{5}$ of the cost of A's horse, is 4 times what he gained by selling it; what was his gain per cent?

23. $\frac{3}{10}$ of the money B paid for silks, is twice what he gained by selling them. What did he gain per cent?

24. $\frac{7}{50}$ of the money C paid for books, is $\frac{1}{5}$ of what he gained by selling them. How much did he gain per cent?

25. $\frac{8}{25}$ of the money I have on interest, is 4 times the yearly interest received. What is the rate per cent?

26. $\frac{13}{45}$ of the cost of A's merchandise, is $\frac{4}{3}$ of what he gained when he sold it. What did he gain per cent?

27. $\frac{4}{75}$ of the cost of B's wagon, is $\frac{2}{3}$ of what he gained by selling it. What did he gain per cent?

28. A book was sold for $\frac{2}{3}$ of $\frac{6}{5}$ of what it cost. What was the loss per cent?

29. $\frac{3}{4}$ of $\frac{8}{5}$ of the cost of a sleigh, was what it was sold for. What was the gain per cent?

30. A, being asked what he gained per cent on a certain stock of goods, replied, that $\frac{3}{4}$ of $\frac{12}{5}$ of what they cost, equals what he sold them for. What did he gain per cent?

LESSON IV.

1. At 5 per cent for 4 years, what part of the principal equals the interest?

SOLUTION 1st.—If the interest on \$1 for 1 year is 5 cents, for 4 years it is 4 times 5 cents, which are 20 cents. If the interest on 100 cents is 20 cents, on 1 cent it is $\frac{1}{100}$ of 20 cents, which is $\frac{20}{100}$, or $\frac{1}{5}$ of a cent. Therefore, at 5 per cent for 4 years, $\frac{1}{5}$ of the principal equals the interest.

2. At 6 per cent for 5 years, what part of the principal equals the interest?
3. At 3 per cent for 2 years, what part of the principal equals the interest?
4. At 4 per cent for 3 years, what part of the principal equals the interest?
5. At 6 per cent for 3 years, what part of the principal equals the interest?
6. At 4 per cent for 3 years, what part of the principal equals the interest?
7. At 9 per cent for 6 years, what part of the principal equals the interest?
8. At 8 per cent for 5 years, what part of the principal equals the interest?
9. At 4 per cent for 6 years, what part of the principal equals the interest?
10. At 6 per cent for 8 years, what part of the principal equals the interest?
11. At 10 per cent for 5 years, what part of the principal equals the interest?
12. At 6 per cent for 4 years and 8 months, what part of the principal equals the interest?

REMARK.—It is expected that pupils will thoroughly understand every lesson that they have been over. Consequently in the following solutions we will merely refer to the lesson (and give the result) by which a particular part of the solution may be performed. But the pupil must go through with the solution as given in the *quoted* lesson, to obtain that result for himself.

SOLUTION 2nd.—We find by *Lesson I.*, that 4 years and 8 months equal $\frac{14}{3}$ years. If the interest on \$1 for 1 year is 6 cents, for $\frac{1}{3}$ of a year it is $\frac{1}{3}$ of 6 cents, which is 2 cents, and for $\frac{14}{3}$ years, 14 times 2 cents, which are 28 cents. If on 100 cents the interest is 28 cents, on 1 cent it is $\frac{1}{100}$ of 28 cents which is $\frac{28}{100}$, or $\frac{7}{25}$ of a cent. Therefore, at 6 per cent for 4 years and 8 months $\frac{7}{25}$ of the principal equals the interest.

13. At 4 per cent for 6 years and 6 months, what part of the principal equals the interest?

14. At 6 per cent for 5 years and 4 months, what part of the principal equals the interest?

15. At $10\frac{1}{2}$ per cent for 1 year and 6 months, what part of the principal equals the interest?

16. At $4\frac{2}{3}$ per cent for 9 years, what part of the principal equals the interest?

17. At $3\frac{3}{5}$ per cent for 2 years and 2 months, what part of the principal equals the interest?

18. At $6\frac{1}{4}$ per cent for 4 months and 24 days, what part of the principal equals the interest?

19. At $7\frac{1}{5}$ per cent for 10 months, what part of the principal equals the interest?

20. At $3\frac{7}{19}$ per cent for 2 years 4 months and 15 davs, what part of the principal equals the interest?

LESSON V.

1. What is the interest of $50 for 4 years, at 6 per cent?

SOLUTION.—We find by *Lesson* IV., that, $\frac{6}{25}$ of the principal equals the interest. $\frac{1}{25}$ of $50 is $2, and $\frac{6}{25}$ is 6 times $2. which are $12, the required interest.

2. What is the interest of $10 for 2 years, at 5 per cent?

3. What is the interest of $48 for 6 years, at 5 per cent?

4. What is the interest of $70 for 7 years, at 5 per cent?

5. What is the interest of $68 for 5 years, at 6 per cent?

6. What is the interest of $70 for 2 years, at 5 per cent?

7. What is the interest of $75 for 5 years, at 3 per cent?

8. What is the interest of $120 for 8 years, at 5 per cent?

9. What is the interest of $100 for 10 years, at 6 per cent?

10 What is the interest of $140 for 12 years, at 5 per cent?

11. What is the interest of $150 for 5 years, at 3 per cent?

12. What is the interest of $145 for 6 years, at 5 per cent?

13. What is the interest of $200 for 10 years, at 8 per cent?

14. What is the interest of $250 for 3 years, at 8 per cent?

15. What is the interest of $220 for 11 years, at 10 per cent?

16. What is the interest of $500 for 9 years, at 8 per cent?

17. What is the interest of $250 for 12 years, at 6 per cent?

18. What is the interest of $500 for 8 years, at 12 per cent?

19. What is the interest of $200 for 9 years, at 3 per cent?

20. What is the interest of $406 for 10 years, at 8 per cent?

21. What is the interest of $50 for 2 years and 2 months, at 6 per cent?

REMARK.—We find (by *Lesson* I,) that, 2 years and 2 months equal $\frac{13}{6}$ years. And, (by *Lesson* IV. *Solution* 2nd.) that, $\frac{13}{100}$ of the principal equals the interest, &c.

22. What is the interest of $25 for 4 years and 3 months, at 4 per cent?

23. What is the interest of $80 for 5 years and 5 months, at 6 per cent?

24. What is the interest of $60 for 8 years and 6 months, at 6 per cent?

25. What is the interest of $240 for 3 years and 9 months, at 6 per cent?

26. What is the interest of $75 for 4 years and 8 months, at 9 per cent?

27. What is the interest of $50 for 2 years and 9 months, at 6 per cent?

28. What is the interest of $80 for 12 years and 10 months, at 6 per cent?

29. What is the interest of $69 for 8 years and 4 months, at 2 per cent?

30. What is the interest of $60 for 4 years and 8 months, at 3 per cent?

31. What is the interest of $600 for 2 years 4 months and 15 days, at 4 per cent?

REMARK.—We find (by *Lesson* I,) that, 2 years 4 months and 15 days equal $\frac{19}{8}$ years. And, (by *Lesson* IV, *Solution* 2nd,) that, $\frac{19}{200}$ of the principal equals the interest, &c.

32. What is the interest of $300 for 5 years 9 months and 18 days, at 5 per cent?

33. What is the interest of $550 for 4 years 7 months and 6 days, at 10 per cent?

34. What is the interest of $500 for 1 year 7 months and 18 days, at 6 per cent?

35. What is the interest of $250 for 2 years 7 months and 6 days, at 4 per cent?

36. What is the interest of $250 for 3 years 3 months and 6 days, at 6 per cent?

37. What is the interest of $50 for 6 years 4 months and 24 days, at 5 per cent?

38. What is the interest of $75 for 2 years 11 months and 6 days, at 15 per cent?

39. What is the interest of $150 for 2 years 6 months and 12 days, at 15 per cent?

40. What is the interest of $300 for 2 years 9 months and 18 days, at $1\frac{1}{3}$ per cent?

LESSON VI.

1. What is the amount of $75 for 2 years, at 6 per cent?

REMARK.—(By *Lesson* IV,) we find, that $\frac{3}{25}$ of the principal equals the interest. The principal, which is $\frac{25}{25}$ added to the interest, which is $\frac{3}{25}$ of the principal, equals $\frac{28}{25}$ of the principal, or the *amount.* Therefore $\frac{28}{25}$ of $75 equals the amount, &c.

Those who prefer, can first find the interest as in the preceding lesson, to which add the principal, and this sum will be the amount.

2. What is the amount of $90 for 3 years, at 7 per cent?

3. What is the amount of $100 for 4 years, at 5 per cent?

4. What is the amount of $160 for 10 years, at 5 per cent?

5. What is the amount of $160 for 8 years, at 5 per cent?

6. What is the amount of $200 for 12 years, at 5 per cent?

7. What is the amount of $210 for 2 years and 6 months, at 4 per cent?

8. What is the amount of $250 for 4 years and 3 months, at 8 per cent?

9. What is the amount of $240 for 4 years and 2 months, at 3 per cent?

10. What is the amount of $500 for 3 years 3 months and 6 days, at 6 per cent?

11. What is the amount of $200 for 5 years 4 months and 24 days, at 5 per cent?

LESSON VII.

1. What principal will in 4 years, at 6 per cent, give $12 interest?

SOLUTION.—We find (by *Lesson* IV,) that $\frac{6}{25}$ of the principal equals the interest. And the interest is $12. Therefore, $12 is $\frac{6}{25}$ of the principal. If $12 is $\frac{6}{25}$ of the principal, $\frac{1}{25}$ of the principal is $\frac{1}{6}$ of $12, which is $2, and $\frac{25}{25}$ (the principal) is 25 times $2, which are $50.

2. What principal will in 6 years, at 4 per cent, give $36 interest?

3. What principal will in 4 years, at 5 per cent, give $30 interest?

4. What principal will in 8 years, at 7 per cent, give $42 interest?

5. What principal will in 10 years, at 7 per cent, give $140 interest?

6. What principal will in 4 years and 6 months, at 6 per cent, give $54 interest?

7. What principal will in 4 years and 3 months, at 6 per cent, give $102 interest?

8. What principal will in 4 years and 3 months, at 8 per cent, give $51 interest?

9. How much money has that man on interest, who, at the expiration of 4 years and 4 months, at 6 per cent, receives $260 interest?

10. At the expiration of 2 years and 4 months, at 6 per cent, a man received $49 interest. How much money had he on interest?

11. A is worth twice as much as B, and the interest of their united fortunes for 4 years and 2 months, at 6 per cent, is $600. How much is each worth?

12. The interest on the cost of B's store and house, for 1 year and 6 months, at 4 per cent, would be $270. What was the cost of each, providing the store cost $\frac{1}{2}$ as much as the house?

13. If the money B paid for a sheep, a cow, and horse, was put on interest for 4 years and 6 months, at 4 per cent, it would give $18 interest. What was the cost of al, and of each respectively, providing the

sheep cost $\frac{1}{3}$ as much as the cow, and the cow, $\frac{1}{4}$ as much as the horse?

LESSON VIII.

1. What principal will in 4 years, at 5 per cent, amount to $360?

SOLUTION.—We find (by *Lesson* IV.) that $\frac{1}{5}$ of the principal equals the interest. This added to $\frac{5}{5}$ (the principal) equals $\frac{6}{5}$ of the principal, or the amount. If $\frac{6}{5}$ of the principal is $360, $\frac{1}{5}$ is $\frac{1}{6}$ of $360, which is $60, and $\frac{5}{5}$ (the principal) is 5 times $60, which are $300.

2. What principal will in 3 years, at 6 per cent, amount to $118?
3. What principal will in 6 years, at 10 per cent, amount to $120?
4. What principal will in 10 years, at 7 per cent, amount to $170?
5. What principal will in 4 years, at 5 per cent, amount to $660?
6. A is worth $\frac{1}{3}$ as much as B; and the interest on their united fortunes for 2 years, at 5 per cent, is $880. What is each of their fortunes?
7. A merchant sold a quantity of cloth for $214, and thereby gained 7 per cent. What did the cloth cost him?
8. What principal will in 2 years, at 7 per cent, amount to $1140?
9. What principal will in 10 years and 8 months, at 9 per cent, amount to $490?
10. The amount due on a note, which had been on interest 6 years and 2 months, at 6 per cent, was $274. What was the face of the note?
11. What principal will in 12 years and 9 months, at 4 per cent, amount to $302?

12. If $\frac{1}{2}$ of A's fortune for 4 years and 6 months, at 6 per cent, amounts to \$127; what is his whole fortune?

13. If $\frac{2}{3}$ of B's fortune, being put on interest for 3 years 3 months and 6 days, at 15 per cent, amount to \$149; what is his whole fortune?

14. A's fortune added to $\frac{2}{3}$ of B's, which is to A's as 2 to 3, being put on interest for 6 years, at 4 per cent, amounts to \$124. What is the fortune of each?

15. D's money added to 4 times E's, which is equal to D's, being on interest for 10 years, at 5 per cent. amounts to \$3000. What was each of their fortunes?

16. The sum of $\frac{2}{3}$ of A's + $\frac{1}{2}$ of B's money, being on interest for 8 years, at 5 per cent, amounts to \$2100. Providing $\frac{1}{2}$ of B's money is twice $\frac{2}{3}$ of A's; how much money has each?

17. $\frac{2}{5}$ of the cost of C's house, increased by $\frac{2}{5}$ of the cost of his farm, being placed on interest for 10 years, at 7 per cent, amounts to \$17000. What was the cost of each, if $\frac{2}{5}$ of the cost of the house was only $\frac{1}{4}$ of $\frac{2}{5}$ of the cost of the farm?

18. If $\frac{5}{7}$ of A's fortune in 2 years and 4 months, at 6 per cent, amounts to \$570; what is his whole fortune?

19. The sum of A's, and B's fortune in 4 years and 8 months, at 6 per cent, amounts to \$256. What was each of their fortunes, providing $\frac{2}{3}$ of A's fortune quals B's?

LESSON IX.

1. In what time will \$40, at 6 per cent, give \$12 interest?

SOLUTION.—What part of the principal equals the interest? $1 is $\frac{1}{40}$ of the principal and $12 is 12 times $\frac{1}{40}$, which are $\frac{12}{40}$ or $\frac{3}{10}$ of the principal. If on 1 cent I gain $\frac{3}{10}$ of a cent, on 100 cents I shall gain 100 times $\frac{3}{10}$, which are $\frac{300}{10}$, or 30 per cent. If it require 1 year for $1 to gain 6 cents interest, to gain 1 cent, it will require $\frac{1}{6}$ of a year, and to gain 30 cents it will require 30 times $\frac{1}{6}$, which are $\frac{30}{6}$ or 5 years.

2. In what time will $60, at 5 per cent, give $18 interest?

3. In what time will $90, at 7 per cent, give $27 interest?

4. In what time will $100, at 6 per cent, give $10 interest?

5. In what time will $120, at 10 per cent, give $120 interest?

6. In what time will $250, at 6 per cent, give $20 interest?

7. In what time will $40, at 7 per cent, give $8.40 interest?

8. In what time, at 8 per cent, will $30 give $9.60 interest?

9. In what time, at 6 per cent, will $10 give $2.40 interest?

10. In what time, at 4 per cent, will $20 give $5.60 interest?

LESSON X.

1. At what per cent, will $50, in 1 year and 6 months, or ($1\frac{1}{2}$ years,) give $6 interest?

SOLUTION.—What part of the principal equals the interest? $1 is $\frac{1}{50}$ of the principal, and $6 is 6 times $\frac{1}{50}$, which are $\frac{6}{50}$ or $\frac{3}{25}$ of the principal. If the interest on 1 cent is $\frac{3}{25}$ of a cent, on 100 cents, it is 100 times $\frac{3}{25}$, which are $\frac{300}{25}$ or 12 cents. If in $\frac{3}{2}$ years the interest on $1 is 12 cents, in $\frac{1}{2}$ of a year it is $\frac{1}{3}$ of 12 cents, which is 4 cents, and in $\frac{2}{2}$ (a year) it is twice 4 cents, wh:ch are 8 cents, or 8 per cent.

2. At what per cent, will $40 annually give $2 interest ?

3. At what per cent, will $80 annually give $3.20 interest ?

4. At what per cent, will $120 annually give $12 interest ?

5. At what per cent, will $120 in 4 years, give $20 interest ?

6. At what per cent, will $100 in 3 years, give $30 interest ?

7. At what per cent, will $5 in 14 years, give $7 interest ?

8. At what per cent, will $25 in 1 year and 9 months, give $3.50 interest ?

9. At what per cent, will $80 in 5 years and 8 months, give $34 interest ?

10. At what per cent, will $500 in 7 years and 6 months, give $75 interest ?

11. At what per cent, will $600 in 2 years 4 months and 15 days, give $57 interest ?

LESSON XI.

1. At what per cent, will $10 in 4 years, amount to $ 2 ?

REMARK.—From the amount, subtract the principal, and the remainder will be the interest. Then proceed as in the preceding lesson.

2. At what per cent, will $12 in 3 years, amount to $13,44 ?

3. At what per cent, will $20 in 6 years, amount to $26 ?

4. At what per cent, will $24 in 10 years, amount to $36 ?

5. At what per cent, will $30 in 7 years, amount to $36.30 ?

6. At what per cent, will $50 in 10 years, amount to $75 ?

7. At what per cent, will $36 in 5 years, amount to $39.60 ?

8. At what per cent, will a given principal double itself, in 20 years ?

SOLUTION.—A given principal will double itself in 1 year, at 100 per cent; and in 20 years, at $\frac{1}{20}$ of 100 per cent, which is 5 per cent ?

9. At what per cent, will a given principal double itself, in 4 years ?

10. At what per cent, will a given principal double itself, in 3 years ?

11. At what per cent, will a given principal double itself, in 5 years ?

12. At what per cent, will a given principal double itself, in 7 years ?

13. At what per cent, will a given principal double itself, in 6 years ?

14. At what per cent, will a given principal double itself, in 8 years ?

15. At what per cent, will a given principal double itself, in 9 years ?

16. At what per cent, will a given principal double itself, in 25 years ?

17. At what per cent, will a given principal double itself, in 30 years ?

18. At what per cent, will a given principal double itself, in $12\frac{1}{2}$ years ?

19. At what per cent, will a given principal double itself, in $14\frac{1}{4}$ years ?

20. At what per cent, will a given principal double itself, in $31\frac{1}{4}$ years ?

LESSON XII.

1. In what time will a given principal double itself, at 5 per cent?

SOLUTION.—A given principal will double itself in 100 years at 1 per cent, and at 5 per cent, in $\frac{1}{5}$ of 100 years, which is 20 years.

2. In what time will a given principal double itself, at 4 per cent?
3. In what time will a given principal double itself, at 3 per cent?
4. In what time will a given principal double itself, at 6 per cent?
5. In what time will a given principal double itself, at 2 per cent?
6. In what time will a given principal double itself, at 7 per cent?
7. In what time will a given principal double itself, at 9 per cent?
8. In what time will a given principal double itself, at 8 per cent?
9. In what time will a given principal double itself, at 10 per cent?
10. In what time will a given principal double itself, at 12 per cent?

LESSON XIII.

1. Bought a bushel of grass-seed, for $5 and sold it for $7; what was the gain per cent?

SOLUTION.—Since it was bought for $5 and sold for $7, the gain must have been $7 — $5, which is $2. If on $5 the gain is $2, on $1 it will be $\frac{2}{5}$ of a dollar, (or $\frac{2}{5}$ of a cent on 1 cent).

If the gain on 1 cent is $\frac{2}{5}$ of a cent, on 100 cents, it is 100 times $\frac{2}{5}$ of a cent, which are $\frac{200}{5}$ or 40 per cent.

2. A book was bought for $2, and sold for $3; what was the gain per cent?

3. A shawl cost $5, and was sold for $8; what was the gain per cent?

4. A cow was bought for $20, and sold for $25; what was the gain per cent?

5. A merchant bought a hogshead of molasses for $80, and sold it for $95; what did he gain per cent?

6. A barrel of pork cost $12, and was sold for $11; what was the loss per cent?

7. A horse was bought for $140, and sold for $60; what was the loss per cent?

8. Bought an orange for 4 cents, and sold it for 6 cents; what was the gain per cent?

9. Bought a melon for 15 cents, and sold it for 20 cents; what was the gain per cent?

10. Bought a book for 5 dimes, and sold it for 8 dimes; what was the gain per cent?

11. Bought a quantity of silk for $120, and sold it for $200; what was the gain per cent?

12. A boy sold melons, at the rate of 10 cents apiece, $\frac{1}{5}$ of which equaled his gain; how much would he have gained per cent, if he had sold them, at 12 cents apiece?

13. A merchant sold sugar, for $80 a hogshead, and thereby cleared $\frac{1}{10}$ of this money; if he had sold it, at $92 a hogshead, what would he have gained per cent?

14. A quantity of cloth was bought for $36, and sold for $43; what was the gain per cent?

15. A horse was bought for $100, and sold for $95; what was the loss per cent?

LESSON XIV.

1. A man bought a cow for $20; for what must he sell her, to gain 5 per cent on the cost?

SOLUTION.—If he gains 5 cents on 100 cents, on 1 cent he will gain $\frac{1}{100}$ of 5 cents, which is $\frac{5}{100}$, or $\frac{1}{20}$ of a cent. Consequently he must sell it for $\frac{21}{20}$ of what it costs, &c.

2. A man bought a yoke of oxen for $100; how must he sell them, to gain 6 per cent on the cost?

3. A man bought a barrel of rum for $10; for what must he sell it, to gain 10 per cent on the cost?

4. A gallon of wine was bought for 20 dimes; how must it be sold a pint, to gain 20 per cent on the cost?

5. A hogshead of molasses cost $20; for what ought it to be sold a gallon, to gain 40 per cent on the cost?

6. B bought a horse for $80, and by selling it, lost 5 per cent on the cost; for what did he sell it?

7. A wagon cost $140 dollars, and was sold for 5 per cent less than it cost; for what was it sold?

8. A merchant, by selling 40 yards of cloth for $164, lost 20 per cent on the cost. What did it cost per yard?

9. If a quart of brandy cost 50 cents, how must it be sold a gill, to lose 4 per cent?

10. B lost 5 per cent, by selling a gallon of rum, which cost 80 cents; for what did he sell it a gallon?

LESSON XV.

1. What principal will, in 4 years, at 5 per cent, amount to $60?

SOLUTION.—We find (by *Lesson* IV,) that $\frac{1}{5}$ of the principal equals the interest. $\frac{5}{5}$ (the principal) and $\frac{1}{5}$ (the interest) added, is $\frac{6}{5}$ of the principal, which equals the amount. If $\frac{6}{5}$ of the principal is \$60, $\frac{1}{5}$ is $\frac{1}{6}$ of \$60, which is \$10; and $\frac{5}{5}$, (the principal,) is 5 times \$10, which are \$50.

2. What principal will, in 3 years, at 6 per cent, amount to \$118?

3. What principal will, in 5 years, at 6 per cent, amount to \$130?

4. What principal will, in 7 years, at 5 per cent, amount to \$81?

5. What principal will, in 9 years, at 8 per cent, amount to \$86?

6. What principal will, in $3\frac{3}{4}$ years, at 8 per cent, amount to \$260?

7. What principal will, in $4\frac{2}{3}$ years, at 6 per cent, amount to \$640?

8. What principal will, in $5\frac{5}{7}$ years, at 7 per cent, amount to \$42?

9. What principal will, in $6\frac{3}{7}$ years, at 7 per cent, amount to \$87?

10. What principal will, in $8\frac{2}{3}$ years, at 6 per cent, amount to \$76?

REMARK.—The *present worth* of a debt, payable at some future time, without interest, is such a sum, as will, in the given time, and at the given rate per cent, amount to the debt. Hence the *present worth* of any sum of money, payable at some future time without interest, may be found in the same way that we found the *principal*, when we had given the amount, time, and rate per cent.

See the above solution.

11. What is the present worth of \$26, due 5 years hence, at 6 per cent? Ans. \$20.

12. What is the present worth of \$14, due 8 years hence, at 5 per cent?

13. What is the present worth of \$110, due 5 years hence, at 5 per cent?

14. What is the present worth of \$86, due 8 years hence, at 9 per cent?

15. What is the present worth of $102, due 9 years hence, at 4 per cent?

16. What is the present worth of $72, due 4 years hence, at 5 per cent?

17. What is the discount on $46, due 3 years hence, t 5 per cent?

REMARK.—*The discount equals the amount minus the present worth.*

18. What is the discount of $54, due 5 years hence, at 7 per cent?

19. What is the discount of $65, due 5 years hence, at 6 per cent?

20. What is the discount of $93, due 3 years hence, at 8 per cent?

21. What is the present worth of $186, due $4\frac{4}{5}$ years hence, at 5 per cent?

22. What is the present worth of $66, due $5\frac{1}{3}$ years hence, at 6 per cent?

23. What is the present worth of $128, due $4\frac{2}{3}$ years hence, at 6 per cent?

LESSON XVI.

1. If I sell cloth, at $2.50 a yard, and thereby gain 25 per cent; what did it cost a yard?

SOLUTION.—25 per cent is $\frac{1}{4}$ of the cost (*Lesson* II.) $\frac{4}{4}$ (the cost) + $\frac{1}{4}$ (the gain) is $\frac{5}{4}$ of what it cost me; which equals what I sold it for. If $\frac{5}{4}$ of what I gave for it, is $2.50, $\frac{1}{4}$ is $\frac{1}{5}$ of $2.50, which is.50 cents; and $\frac{4}{4}$ (the cost) is 4 times 50 cents, which are 200 cents, ($2.)

2. A horse was sold for $38, which was at a loss of 5 per cent. What did the horse cost?

3. If I sell cloth, at $2.50 a yard, and thereby gain 25 per cent, how must I sell it a yard, to lose 20 per cent?

4. If I sell cloth, at $4.40 a yard, and thereby gain 10 per cent, how ought I to sell it, to lose 25 per cent?

5. If by selling a piece of cloth for $46, I gain 15 per cent, how ought I to have sold it, to have lost 30 per cent?

6. A sold his horse for $105, and thereby gained 5 per cent on the cost; for what ought he to have sold it, to have lost 10 per cent?

7. A farm was sold for $495, which was 10 per cent less than what it was worth; for what ought it to have been sold, to have received 40 per cent more than its value?

8. A mechanic lost 20 per cent on the cost of a wagon, by selling it for $40; for what ought it to have been sold, to have gained 30 per cent?

9. A horse was sold for $90, which was 10 per cent less than its value; what would have been the gain per cent, if it had been sold for $120?

10. A farm was sold for $690, which was 8 per cent less than its value; what would have been the gain per cent, if it had been sold for $850?

11. A book was sold for 90 cents, which was 10 per cent less than its value; what would have been the gain per cent, if it had been sold for $1.50?

12. A man sold two watches, at $12 each; on one he gained 50 per cent, and on the other he lost 50 per cent. Did he gain or lose by the bargain, and how much?

13. An individual sold two gold pencils, at $6 apiece; on one he gained 20 per cent, and on the other he lost 20 per cent. Did he gain or lose, and how much?

14. A farmer sold two horses at $210 apiece; for one he received 25 per cent more than its value, and for the other 25 per cent less than its value. Did he gain or lose by the bargain, and how much?

15. A merchant sold a quantity of cloth for $80, and by so doing lost 60 per cent; he then sold another quantity for $80, and thereby gained 60 per cent. Did he gain or lose by the operation and how much?

LESSON XVII.

1. An individual was ordered to collect $190, and was to receive 5 per cent on all the money collected. How much should he receive?

REMARK.—It will be understood that he is to collect his own fee, and must receive 5 per cent on that also.

SOLUTION.—We find (by Lesson II.) that he is to receive $\frac{1}{20}$ of all he collects. Therefore $\frac{20}{20}$, (what he collects)—$\frac{1}{20}$ (his fee) is $\frac{19}{20}$ of what he collects, which is $190, (the sum to be paid to his employer.) If $\frac{19}{20}$ of what he collects equals $190, $\frac{1}{20}$ is $\frac{1}{19}$ of $190, which is $10; and $\frac{20}{20}$ (what he collects) is 20 times $10, which are $200. Therefore he must receive $200—$190=$10.

2. What ought A to receive for collecting $90, if he receive 10 per cent on all he collects?

3. What amount of money will be sufficient to pay a debt of $38 and the collector's fee, which is 5 per cent on all the money collected?

4. How much cider must that man make to fetch away 15 barrels, after the owner of the mill receives $16\frac{2}{3}$ per cent of all he has made?

5. How much grain must a farmer take to mill, that he may bring away the flour of 1 bushel, after the miller has taken 10 per cent. of all he took there?

MISCELLANEOUS EXAMPLES.

1. At 5 per cent for 4 years, what part of the principal equals the interest?

2. In how many years, at 4 per cent, will a given principal amount to the same, as it would in 8 years, at 6 per cent?

3. At what per cent will a given principal, in 14 years, amount to the same, as it would in 12 years, at 7 per cent?

4. If $\frac{1}{10}$ of the principal equals the interest, what is the rate per cent?

5. The rent of B's farm, for 8 years amounted to $\frac{12}{10}$ of its value. What per cent did he annually receive on the value of his farm?

6. What is the interest of $75, for $5\frac{3}{5}$ years, at 5 per cent?

7. What principal will, in $7\frac{1}{2}$ years, at 8 per cent, give $24 interest?

8. What principal will, in $4\frac{4}{5}$ years, at 5 per cent, amount to $155?

9. At what per cent, will a given principal double itself, in $12\frac{1}{2}$ years?

10. The interest of A's, and B's fortune, for 8 years, at 5 per cent, is $420. What is the fortune of each, providing A's fortune is twice B's?

11. The interest of $\frac{2}{3}$ of A's and $\frac{3}{4}$ of B's fortune, for 7 years, at 5 per cent, is $2100. What is each of their fortunes, providing $\frac{2}{3}$ of A's fortune equals $\frac{3}{4}$ of B's?

12. B sold his horse, for $\frac{1}{2}$ of $1\frac{1}{2}$ times what it cost; what did he lose per cent?

13. What is the interest of $540, for 4 years, at 5 per cent?

14. What is the interest of $180, for 5 years and 9 months, at $6\frac{2}{3}$ per cent?

15. What principal will, in 4 years 7 months and 8 days, at $6\frac{1}{4}$ per cent, amount to $412?

16. The interest of the cost of B's horse, sleigh and wagon, for 6 years, at 5 per cent, is $69. What is the cost of each, providing their prices are to each other respectively, as $\frac{1}{2}$, $\frac{2}{3}$ and $\frac{3}{4}$?

17. What principal will, in 8 years and 8 months, at $7\frac{1}{5}$ per cent, amount to $419?

18. What principal will, in 5 years 9 months and 18 days, at 10 per cent, give $116 interest?

19. In what time, will $420, at 5 per cent, give $147 interest?

20. If the interest of $200, for 1 year and 6 months, is $18, what is the rate per cent?

21. At what per cent, will $500, in 4 years and 9 months, give $190 interest?

22. At what per cent, will $500, in 22 years and 6 days, amount to $1821?

23. At what per cent will a given principal double itself, in 20 years?

24. In what time will a given principal double itself, at 12½ per cent?

25. At what per cent will a given principal double itself, in 6 years and 8 months?

26. A horse was bought for $60, and sold for $90; what was the gain per cent?

27. A basket containing 39 oranges, cost $1.20; how must they be sold apiece to gain 30 per cent?

28. If 1 quart of champagne cost 40 cents, how must it be sold a gill to gain 20 per cent?

29. What is the present worth of $68, due 10 years hence, at 7 per cent?

30. What is the discount on $162, due 10 years and 4 months hence, at 6 per cent?

31. What is the present worth of $87, due 3⅕ years hence, at 5 per cent?

32. If a hogshead of molasses containing 84 gallons cost $30, how must it be sold a gallon, to gain 40 per cent?

33. The money I have on interest, in 9 years, at 10 per cent, amounts to $190; what is the principal?

34. When money was worth 6 per cent, I bought $400 worth of goods; 6 months afterwards I sold them, and gained 10 per cent on the cost. How much did I gain? Ans. $28.

35. A speculator bought a horse for $36, and sold it for 25 per cent more than he gave for it; which, however, was 10 per cent less than what

he asked for it. How much did he ask for the horse ?

36. A gentleman being asked how much money he had on interest, replied, that if instead of 6 per cent. he should receive 10 per cent., he would receive $268 interest more than he then did. How much money had he on interest ?

37. A merchant bought broadcloth for $1.20 a yard and sold it for 33$\frac{1}{3}$ per cent. more than he gave for it; which, however, was 33$\frac{1}{6}$ per cent. less than his marked price for it. How much was his marked price per yard ?

38. A merchant sold a quantity of cloth for $120, and by so doing, gained 50 per cent. He then sold another quantity, for $120, and thereby lost 50 per cent. Did he gain or lose by the bargain, and how much ?

39. B sold a horse for $60, and gained 20 per cent. He then sold another horse for $60, and lost 60 per cent. Did he gain or lose, and how much ?

40 The interest on 1$\frac{1}{3}$ times A's, and $\frac{2}{5}$ of B's fortune, for 8 years, at 5 per cent. is $520. What is the fortune of each, providing 1$\frac{1}{3}$ times A's fortune, equals $\frac{2}{5}$ of B's ?

41. $\frac{2}{3}$ of D's fortune added to $\frac{3}{4}$ of E's, which is 3 times $\frac{2}{3}$ of D's, being put on interest for 8 years, at 5 per cent. gives $800 interest. What is the fortune of each ?

42. The interest of A's, B's, and C's fortune, for 5 years, at 8 per cent. is $1040. What is the fortune of each, providing they are to each other as $\frac{1}{2}$, $\frac{1}{3}$, and $\frac{1}{4}$?

43. The interest of A's, B's, and C's fortune, for 5$\frac{1}{3}$ years, at 6 per cent. is $800. What is each of their fortunes, providing B's, is twice A's; and B's, and C's, are equal ?

44. The interest of the sum of $\frac{1}{2}$ of A's, and $\frac{2}{3}$ of

B's fortune, for a *certain* time, at 2 per cent. was to *this sum* as 9 to 250. And the amount of *this interest* for 25 times as long, at 10 times as great a per cent was $180. What was each of their fortunes, providing A's fortune was to B's as 1 to 3? And how long was the first on interest?

REMARK.—Since the interest was to the principal as 9 to 250, $\frac{9}{250}$ of the principal equals the interest. By *Lesson IX*, this readily gives 1 year 9 months and 18 days for the required time.—By the first condition of the question $\frac{9}{250}$ of the principal equals the interest; and for 25 times as long, at 10 times as great a per cent. the interest would be 9 times the principal; which added to 1 (the principal) is 10 times the principal (the first interest,) which equals $180 (the amount.) Therefore $18 was the principal (the first interest,) which equaled $\frac{9}{250}$ of the first principal, which we readily find to be $500, which is the sum of $\frac{1}{2}$ of A's and $\frac{2}{3}$ of B's fortune. And, since A's fortune was to B's as 1 to 3, we easily find that $\frac{2}{3}$ of B's fortune equals 4 times $\frac{1}{2}$ of A's. Consequently $\frac{1}{5}$ of $500 was $\frac{1}{2}$ of A's fortune, and $\frac{4}{5}$ of $500 was $\frac{2}{3}$ of B's. Therefore, A's fortune was $200, and B's $600.

RECOMMENDATIONS

From the State of New York.

Secretary's Office, ALBANY, Dec. 15, 1849. *Dep. of Com. Schools*

I have examined "Stoddard's American Intellectual Arithmetic, and cheerfully recommend it to Teachers and Parents, as a valuable elementary work, and one well adapted to the wants of pupils in the first stages of Arithmetic. It is constructed upon sound and practical principles, and will be found an important addition to the text-books now in use in our Common Schools.

SAMUEL S. RANDALL,
Deputy Sup't, Common Schools.

GENEVA, N. Y., Oct. 8th, 1850.

Having examined "Stoddard's Intellectual Arithmetic," I am satisfied it is a work of superior merit, and have concluded to introduce it into our school.

J. SWIFT,
Principal of Geneva Union School.

FROM ELMIRA.

From a knowledge of the success of teaching from John F. Stoddard's "American Intellectual Arithmetic," in the New York State Normal School, and the Albany Male Academy, under the charge of the Rev. Dr. Campbell, I have no hesitation in saying that it is EMINENTLY SUPERIOR TO COLBURN'S Mental Arithmetic. To speak of all its peculiar merits I deem wholly unnecessary on the present occasion; but would simply remark, that it is more Elementary than Colburn's, and embraces a simplification of all the rules of common Arithmetic, as well as many of the more abstruse principles of the higher Mathematics. I have purchased a supply for introduction.

A. R. WRIGHT, Principal.

Elmira Union School, No 2; Sept. 18, 1850.

FROM BUFFALO.

The undersigned Principals of Public Schools in the city of Buffalo, N. Y.,—(the former having carefully examined, and the latter having thoroughly *tested* in his School-Room,)—"Stoddard's Intellectual Arithmetic," take pleasure in saying that it is superior to Colburn's, for the following reasons: It is systematically and Philosophically arranged, contains a much greater variety of Classes of Practical Examples, is adapted to the *currency* of our country, embraces an *original* and very important chapter on the subject of Interest, Discount, and Per-Cent., and avoids those sudden changes from simple to complex Examples, which is the most objectionable feature in Colburn's; and withal, is published in a superior style.

J. A. DOLE, Principal of No. 1,
and Pres't. of Buffalo Teachers Association.
W. W. NEWMAN, Principal of No. 13,
and Sec'y. of Buffalo Teachers Association.

RECOMMENDATIONS.

LIBERTY, N. Y., Jan. 1st, 1850.

Mr. Stoddard is a practical teacher and his "American Intellectual Arithmetic" is the result of much labor, and its publication was suggested by its great advantages in his own experience as an Instructor. I have examined the work with considerable care, and great pleasure; and I have no hesitation in saying that it is admirably adapted to promote mental discipline, and the study of numbers, both in their simple and more complicated forms. The arrangement is so natural, and the transition from the most simple to the most abstruse questions so easy and gradual (I might say imperceptible), that the difficulties of Arithmetic have vanished, and habits of mental exertion are formed before the pupil is aware of it. I am happy to learn that the book is already extensively introduced into schools, as a standard text-book. From its superiority to any of the same kind, with which I am acquainted, and having personally observed the astonishing progress made by those who have studied it under Mr. Stoddard's own training, I can with the utmost confidence recommend it as a class-book, to all who have the instruction of children and youth in schools or academies.

JAMES PETRIE, A. M.,
Professor of Languages in Liberty Normal Institute.

Extract from the Report of the Committee on Text-Books of the Onondaga County Teachers Institute. Adopted April, 1850.

"Colburn's Mental Arithmetic" is generally considered an original and standard work; indeed, almost beyond the hope of competition or improvement. The late DAVID P. PAGE, says of it, "Colburn's was the first, and it is probably the best that has been prepared. That little book has done more than any other for the improvement of teaching in this country." Mr. Page, however, often spoke of its imperfections, and intended to prepare a work himself on this subject. His sudden death deprived the public of that important educational auxiliary; but a volume has been prepared by Mr. John F. Stoddard, a graduate of the New York State Normal School, which your Committee venture to recommend to be substituted for Colburn in all our schools. Two of your Committee, some months since, became so favorably impressed by an examination of this work, as to try the experiment of its use in their schools; and they very freely say, that though Colburn's has long been with them a favorite and much-used volume, they are fully satisfied, by trial in the school-room, that Stoddard's is very much its superior.

Of Colburn's work, we will briefly say, that it is not sufficiently gradual and systematic; there are too many important omissions; too many sudden transitions from examples that are very easy, to those that are very difficult: it is not sufficiently extended in some parts: it is not well adapted to the currency of our country, and there is not an appropriate variety of classes of questions. These deficiencies, we believe, are obviated by the labors of Mr. Stoddard. The first half of his book is on a similar, but *improved plan* of Colburn's. Some portions of the last half are much more difficult than Colburns; and it includes many classes of examples, not introduced into any other mental arithmetic yet published. It embraces some fifty pages of examples suitable for algebraic as well as arithmetical solutions, and about the same amount on per centage, in all its variations, on an entirely new system, and of an original, philosophic, and practical application. These two portions, at least for advanced pupils, we think entitle "Stoddard's American Intellectual Arithmetic" to a more favorable consideration than any other work now before the public; and we therefore cordially commend it to the favorable notice of our teachers, confidently believing, that its superior merits will ultimately banish the very popular, though far from being perfect, work of Colburn from all our schools.

JAMES JOHONNOT,
H. G. McGONEGAL,
A. BRADBURY,
E C. PALMER, } *Committee.*

Report and Book were adopted by the Institute.

Messrs. Cornish, Lamport & Co.:

GENTS:—I have, with considerable care, examined a copy of the "American Intellectual Arithmetic," by John F. Stoddard, and take pleasure in announcing that I regard it as superior to any other work of the kind with which I am acquainted. I think it excels both Colburn's and Thompson's Mental Arithmetic, in presenting as it does, all their simplicity, and at the same time, a more systematic and extended view of mental operations. It is eminently calculated to invigorate and develop the reasoning faculties, and prepare the pupil to acquire, comparatively with ease, a thorough knowledge of mathematics.

WM. HUMPHREY,
1st Assistant Teacher in the Union School, Geneva, N. Y.

October 1850

MIDDLETOWN, N. Y., Aug. 26, 1850.

GENTS:—Having examined "Stoddard's Juvenile and Intellectual Arithmetics," I freely state that I deem them the best series of the kind I have ever met with, and well adapted to properly discipline and develop the minds of pupils. I have already introduced them in the schools of this village and in the Wallkill Academy, and, as opportunity presents, will introduce them in the schools throughout the town.

Yours, &c. H. EVERETT,
Town Superintendent.

PORT JARVIS, N. Y., Aug. 27th, 1850.

From a careful examination given to "Stoddard's Introductory Lessons and Mental Exercises in Intellectual Arithmetic," I feel assured that the public are under obligations to the Author, for the happy manner in which he has graduated the science to the capacity of the juvenile mind.

Mr. Stoddard seems to keep simplicity in view, and at the same time aims to enable the Student to arrive at a thorough understanding of this most indispensible Science. I have purchased a quantity of the Arithmetics from your Agent, for immediate introduction into the schools of this town.

Yours truly, ALEXR. T. JOHNSON,
Town Superintendent, &c.

GOSHEN, ORANGE CO., N. Y., Aug. 21, 1850.

GENTS:—Having examined "Stoddard's Intellectual Arithmetic" published by you, we cheerfully recommend it to the favorable consideration of Teachers, as being superior to anything of the kind we have yet seen. It has been introduced into the schools in this village, and we trust that by its merits it will soon gain admission into the Schools and Academies of the State.

CHAS. T. JANSEN, Town Superintendent.
J. H. THOMPSON, Teacher.
J. C. WALLEN, SAMUEL R. OWEN, N. W. BODLE, } Trustees of the Village of Goshen.

From the Journal of Education and Teachers Advocate, Albany.

This Book is an improvement on the Peztalozian and Colburn plan of teaching, and ought to be in all the Elementary Schools of our country It treats practically of the Elementary Rules, of Fractions, of Interest, Discount, and Per Centage, in a way that cannot fail to engage attention.

WILLIAMSBURGH, Sept. 2d, 1850.

GENTS :—Having used Professor John F. Stoddard's "Primary and Intellectual Arithmetics" in my school for the last twelve months, with complete success, I take great pleasure in recommending them to Teachers, and all others interested in the educational improvement of Youth, as the most perfect work of the kind now in use.—And do not hesitate to say that pupils can by the study of this Arithmetic, make more rapid progress in acquiring a thorough knowledge of numbers, than by any other course. I challenge the Schools of the United States to produce an example of greater proficiency in Mental Arithmetic, than my class of boys from 8 to 12 years of age can furnish. Hoping that so valuable an addition to the text-books of the present day may soon be found in all the Schools of our country—

I remain very respectfully yours, &c.
WILLIAM H. BUTLER,
Principal of Public School, No. 1, Williamsburgh, L. I.

TO THE PUBLISHERS OF "STODDARD'S MENTAL ARITHMETIC."

SIRS :—Your agent has placed the above work in my hands, and I am so well pleased with what I have seen of it, that I have concluded to give it a trial in my Schools, and have taken forty copies for introduction.

M. R. BARNARD,
Principal of Lancasterian School, and Town Sup't. Ithaca.
October 3d, 1850.

PHELPS, N. Y., October 10, 1850.

This is to certify that I have examined "Stoddard's Mental Arithmetic," and am well pleased with it, so much so, that I have introduced into the Vienna Union School as a text-book.

LEWIS PECK, Principal.

www.ingramcontent.com/pod-product-compliance
Lightning Source LLC
LaVergne TN
LVHW021359110826
845150LV00007B/1716

* 9 7 8 1 4 2 5 5 1 3 3 2 0 *